T0269004

SIR THOMAS BROWNE
RELIGIO MEDICI

EDITED AND ANNOTATED BY
JAMES WINNY

CAMBRIDGE
AT THE UNIVERSITY PRESS
1963

CAMBRIDGE UNIVERSITY PRESS
Cambridge, New York, Melbourne, Madrid, Cape Town, Singapore, São Paulo, Delhi

Cambridge University Press
The Edinburgh Building, Cambridge CB2 8RU, UK

Published in the United States of America by Cambridge University Press, New York

www.cambridge.org
Information on this title: www.cambridge.org/9780521043502

First published 1963
This digitally printed version 2008

A catalogue record for this publication is available from the British Library

ISBN 978-0-521-04350-2 hardback
ISBN 978-0-521-09055-1 paperback

FOR

Tim

CONTENTS

The figures on p. 41 illustrating the story of Habakkuk are reproduced from the series of woodcuts after Holbein first published in *Historiarum veteris testamenti icones* (Trechsel, Lyons, 1538) and later reprinted in many bibles.

ACKNOWLEDGEMENTS

Several of the notes to this edition have been written with the guidance of colleagues and friends who have allowed me to consult their special knowledge or wiser judgement. I am grateful to each of them. The introduction owes most to its first reader, who persisted in criticism until its most serious shortcomings had been corrected. This formal acknowledgement does not suggest for how much patience and discernment I am expressing thanks.

J. W.

NOTE ON THE TEXT

This modernised edition of *Religio Medici* is based upon the text established by J.-J. Denonain and published by the Cambridge University Press in 1955. The edition produced by Vittoria Sanna, published at Cagliari in 1958, has been consulted where points of textual obscurity were met. I have departed from the reading of Denonain's text at the following points:

p. 15, l. 26   are but one in him [for, 'are not one in him'].

p. 28, ll. 6–7   not only birds but...beasts [for, 'not onely beasts, but...birds'].

p. 29, l. 18   the doubtful familist's [for, 'the doubtfull Famulies'].

p. 38, l. 7   the fire and scintillation [for, 'and scintillation'].

    l. 16   without this, to me [for, 'to me, without this'].

p. 40, ll. 11–12   so theirs a faculty [for, 'so there's a Faculty'].

p. 42, ll. 21–2   to have left no description [for, 'to have left description'].

p. 48, ll. 21–2   exaltation of gold [for, 'exaltation of God'].

p. 80, ll. 2–3   if not invisible [for, 'if not divisible'].

Like Sanna, I find the reading of p. 48, ll. 10–11 unintelligible, ['ourselves being not yet without life, sense and reason'], but cannot propose any acceptable emendation.

INTRODUCTION

Browne's motive in setting down the private ruminations and disclosures that form *Religio Medici* has not been satisfactorily explained. In the preface to the first authorised edition of the book he affirms that he had intended nothing more than a memorandum of his reflections, addressed to himself; and implies that only the action of Andrew Cooke, who issued a pirated text of the work in 1642, forced him to acquiesce in its publication. We are not obliged to regard this reluctance very seriously. The suggestion that *Religio Medici* was meant to remain a private confession goes no further than the preface, which Browne added to his book after he had suffered the embarrassment of public criticism. We may acknowledge that he might have been taking stock of his personal position at the end of a period of decisive experience, and that *Religio Medici* was written to crystallise an attitude of mind which had been gradually resolving itself. To this extent Browne's purpose was as private as he affirms. But the general style of *Religio Medici* is not that of a private memorandum, nor does Browne's manner suggest meditative self-analysis. Taken as a whole, the book provides a vigorously self-assertive display of Browne's idiosyncrasies, directed towards an audience capable of appreciating both Browne's erudition and his developed taste for outlandish ideas.

What we know of Browne, partly by his own admission, suggests that except in his writing he was retiring and un-communicative. The dramatic disclosures of his inner life offered in *Religio Medici* seem to represent a compensating activity, designed to offset the sense of personal inadequacy which his withdrawn and unsociable nature encouraged. To have revealed this colourful inner self to a circle of close friends,

while remaining in the dark to the world, might have gratified the sense of mystery which Browne draws about his
individuality in his book. Formal publication of *Religio
Medici* thrust upon this shy man the vexation of seeing the
fantasy of his inner life dissected by an amused critic. It
gave Browne a cogent reason for resenting Andrew Cooke's
presumption, and for disclaiming relationship with his younger
self.

Could Browne have foreseen the popularity which his book
was to enjoy, he might have been less anxious to dissociate
himself from what he had written seven years earlier. We may
suspect that Browne—too excited by the discovery of unfathomable curiosity in himself to be soberly critical—revealed
more of his essential simplicity of mind than he realised. But
his lack of critical self-awareness has its own attractions. The
picture of the author which emerges from *Religio Medici* has
an immediate human appeal, enhanced rather than qualified
by the histrionic disclosures and inconsistencies of thought
which betray Browne's designs upon the reader. The pride
which he repeatedly shows in his remarkable individual
qualities, and his patronising references to the less happily
gifted mass of mankind, show the workings of an impulse to
compensate for the actual shyness and reticence of his nature.
It is a paradox such as Browne might have enjoyed, that a
mind so resolutely conservative in its beliefs should be so
concerned to prove itself unaccountable; the culminating
oddity of a world crammed with bizarre creatures and circumstances. The individuality which he claims is so obviously
synthetic—a compilation of attributes drawn from Patristic
and Renaissance writers as well as from reflection upon his
own behaviour—that his personal pride cannot be seriously
regarded. The thought of *Religio Medici* is characterised by
its underlying playfulness.

Had Browne's powers of self-analysis been greater he might

have reflected more purposefully upon his motives in setting down this record of his private ruminations. His declaration that *Religio Medici* was written for his private exercise and satisfaction is useful only for its hint that Browne expected to attract respect rather than notoriety by his disclosure of odd thoughts within conventional attitudes; of conformity uncompromised by a taste for the bizarre. This unexamined purpose still invites speculation. Of the immediate circumstances of Browne's life in 1635, perhaps the most significant is that he had newly returned to England after two years' residence and study abroad. The experience is likely to have presented a challenge to the religious and philosophical orthodoxy which *Religio Medici* proclaims. Both at Montpellier and at Padua, where Browne augmented his medical studies between 1631 and 1633, new currents of thought were stirring. At Montpellier, science and divinity were treated as separate subjects, so that scientists were free to pursue enquiry into the nature of the physical world without being obliged to relate their discoveries to a final cause in God's purpose. Padua afforded still greater academic liberty. Its medical school was famous for its teaching of anatomy, a study commonly impeded by a long-standing prejudice against dismembering the human body; and the university was permeated by a freedom of thought which tolerated, and perhaps encouraged, scepticism and agnosticism. Browne's earlier experience as a medical student at Oxford could have provided no contact with unorthodoxy of this kind. How he responded to the challenge of such revolutionary ideas he does not tell us directly, though he remarks of the heretical writers whom he has read that he could discover 'nothing that may startle a discreet belief'. The nonchalance of the statement, so typical of Browne's pride in the breadth and maturity of his experience, probably reflects a genuine immunity to progressive ideas. His admission of being moved to tears by a religious procession while his companions mocked suggests that he did not adopt the fashionable scepticism of his new environment.

It is hard to believe that an alert and active mind could have remained unaffected by the vigorous intellectual life of these two university cities; but if the conservatism which *Religio Medici* reveals was already a dominant feature of Browne's outlook, he was well protected against the attractions of the new thought. The amused contempt which he expresses for the wilder hypotheses both of the old and of the new science— 'Some have held that snow is black, that the earth moves'— implies a genuine sense of superiority, and not the resentment of a mind which sees its familiar supports being undermined. *Religio Medici* encourages us to assume that Browne was essentially unaffected by contact with the new intellectual forces: that he was perhaps incapable of appreciating the significance of the growing trend away from Scholasticism towards a more pertinent form of scientific enquiry, or of recognising its implicit threat to his world of belief. If it served only to confirm his faith in the authority of the old philosophy, Browne's residence abroad must be reckoned a decisive experience, whose influence was still working upon him when, within a year of returning to England, he set down the 'memorial' which Andrew Cooke published surreptitiously in 1642.

On this simple assessment, *Religio Medici* amounts to a vigorous reaffirmation of conservative and authoritarian principles after the temporary shock of a challenge from the new world of ideas. Contact with progressive thought did not widen Browne's intellectual horizon but drove him in the reverse direction, towards the inert formulas of Scholastic belief which led all speculation back to the contemplation of immutable order. As its title suggests, *Religio Medici* is a review of Browne's religious beliefs, and of the attitudes towards science and philosophy which those beliefs determine. Such a review was bound to involve some measure of self-analysis, and in fact Browne's book becomes increasingly concerned with his private idiosyncrasies and reflections upon a number of topics, not always related to religion. Its introductory pages discuss his open-mindedness towards other forms of Christian faith, and with the peculiarities of belief which he allows himself. After a short consideration of Hermetic philosophy he then moves to the subject which dominates *Religio Medici* as central topic, 'the cosmography of myself'. His absorption in this subject, 'that bold and adventurous piece of nature', might suggest the kind of intellectual curiosity which was to have its consequence in empirical science; but Browne's studies of himself do not tend towards a clearer understanding of his physiological being. Instead, they merely develop the familiar analogy between individual man and the macrocosm. Again, when he announces that his reflections in divinity have two sources, the Bible and 'that universal and public manuscript' of the natural world, he seems to be declaring an attentive interest in his material environment; but Browne shows a persistent disinclination, or inability, to consider the scientific significance of what he observes. The objects which he contemplates in the material world offer him no hints of the natural principles which sustain them, but appear to him as 'mystical types' or emblems of the spiritual world which the material dimly reflects. His self-examination and his religious belief are thus closely associated. What he discovers within himself he identifies with a corresponding feature of the natural world; and what he observes about him he interprets as a shadowing of spiritual reality, where the natural object stands as mystic emblem. In these respects Browne separates himself from the current of scientific curiosity which he had certainly encountered abroad.

When he wrote *Religio Medici* Browne was thirty years old and still unmarried. The book leaves the impression of a much older writer. Many of its sentiments are those of saddened middle age, conveyed with a resignation that is at odds with Browne's actual state of life and genial temper. The curious wish that mankind might reproduce itself like trees, without 'this trivial and vulgar way of coition', is one of many dry and dispirited confessions which Browne seems to be trying out, as a means of catching his reader's fancy. Another characteristic remark, that age 'doth not rectify but incurvate our natures, turning bad dispositions into worser habits', shows Browne assuming an attitude difficult to reconcile with his comparative youthfulness, or with his exuberant manner elsewhere. By adopting attitudes not always consistent with one another, and often at variance with his general eagerness to display himself and his ideas, Browne throws suspicion upon the central personality of *Religio Medici*. What he presents is not a self-portrait but a composite figure, basically himself, but enveloped in borrowed attributes which magnify and enhance his individuality. Montaigne, with whose essays he seems to have been acquainted, could have warned him that 'Il n'est pas description pareille en difficulté à la description de soy-mesmes'; but Browne was not attempting to examine himself with the critical discernment which Montaigne had brought to the task. He sees himself as a uniquely individual being, but also as a representative man in whom the attributes and characteristic experiences of a species are stored. Browne cannot overlook the formal dignity of his position at the centre of the universal stage, or the moral drama of man's divided nature in which he finds himself involved. He presents to his readers a portrait of man who is master of creation, yet humbled by a persistent moral frailty; a being whose ranging intelligence is confounded by the unfathomable wisdom of his Maker. Browne places himself at the centre of this universal concept; a private individuality bearing a cosmic significance and illustrating the moral tradition which saw in man the most splendid and the most tragic of God's creatures. This tacit purpose encourages Browne to attribute to himself the main features which late Renaissance thought had assigned to man, and to lay claim to them as personal reflections and experiences.

His absorption in the little world of himself is justified by Browne's sense of man's importance in the cosmic design, and is not a tribute paid to his own remarkable uniqueness. But so far as *Religio Medici* is offered as private confession, its persistent heightening and straining of sensation must obstruct Browne's purpose by substituting rhetorical flourish for the unaffected speaking voice. When he asserts, 'Lucifer keeps his court in my breast; Legion is revived in me', we may reasonably look for evidence of his possession in a disturbed literary style; a sense of nervous strain and unrest communicated through the structure of his prose. Instead, we find a settled balance of expression that is incompatible with spiritual conflict. Such bold gestures may help to realise Browne's conception of himself as the focus of a universal moral conflict, but they leave us still unacquainted with the inner personality from which *Religio Medici* has emerged.

The curious disparity of tone between Browne's preface and the body of *Religio Medici* is paralleled in the contradictions of his autobiography. Montaigne allows his readers to see the untidiness and occasional banality of thought that he recognises in himself, but Browne offers a picture carefully pruned and trimmed of the irregularities which all human character contains. At the same time, by insisting upon the oddity of his mind, he attempts to invest his character with more mystery than the disclosure of some unaccountable lapse might provide. 'There is, I think,' he remarks with typical complacence, 'no man that apprehends his own miseries less than myself, and no man that so nearly apprehends others.' Yet earlier he has claimed for himself, 'I think no man ever desired life as I have sometimes death'. His object is more frequently striking rhetorical effect rather than perceptive self-analysis; and in the service of this greater interest he is prepared to modify, and sometimes even to invert what he knows of himself. One of the recurrent claims which he seems eager to have us credit is the impartiality of taste and temperament which enables him to participate in all the varieties of human experience. 'I am of constitution so general,' he remarks, 'that it consorts and sympathiseth with all things. I have no antipathy, or rather idiosyncrasy, in diet, humour, air, anything.' In fact, Browne exhibits some marked prejudices, of which perhaps the most conspicuous is his addiction to the odd and the contradictory in preference to unambiguous fact. His readiness to sympathise with Roman Catholic ritual, though himself a declared Anglican, supports his claim to impartiality; but his references to the Copernican theory—perhaps the only novel idea considered by Browne which has a genuine scientific interest—show a mind 'blind with opposition and prejudice' such as he attributes to religious scepticism.

His claim for a universality of interest seems to conflict with the shyness and unsociability which he admits just as frankly. 'I am no way facetious,' he confesses, meaning gay or witty, 'nor disposed for the mirth and galliardise of company.' The general evidence of *Religio Medici*, with its picture of a mind turned inwards upon its own curious habits, makes this claim appear plausible; and Browne's admission of his lack of affection for his parents in Part Two is psychologically compatible with the withdrawn and unsociable nature which he confesses. But when he goes a stage further, and tells us that 'in my retired imaginations I cannot withold my hands from violence on myself', he strains credulity too much. We are likely to prefer his confession of respect for commonsense expediency, 'From...the natural respect that I tender unto the

conservation of my essence and being, I would not perish
upon a ceremony, politic point or indifferency'; not simply
because this lacks the grandeur which Browne too willingly
claims for his actions, but because such a readiness to temporise
is more akin to the unheroic attitude of common humanity.
The timidity and reserve which Browne acknowledges in himself are too much at variance with the dramatic assertions of
Religio Medici for both to be accepted as authentic expressions
of one character. By investing his human condition and ideas
with an arresting significance, Browne seems to be trying to
distract his attention from the admitted inadequacy of his
social self, and elaborating the impressive figure which is
projected throughout *Religio Medici*. 'Now, for my life, it is
a miracle of thirty years, which to relate were not an history
but a piece of poetry, and would sound to common ears like
a fable. For the world, I count it not an inn but an hospital,
and a place not to live in but to die in.' Nothing that we know
of Browne's early life justifies this typically grandiose assertion.
It is not necessarily false; but the excitement which the passage
conveys springs from Browne's adopting of two fundamentally
different attitudes to life; one exulting, the other depressed.
He may persuade us to share his pleasure by identifying ourselves with his excitement, but we must see that the link be-
between the two sentences is merely rhetorical, the second
sharing none of the ideas of the first and inverting its spirit of
astonished elation. The casual association of the two remarks
underlines the impartiality of Browne's taste for oddity, and
his eagerness to annex such ideas. Our readiness to credit his
claim to represent 'a miracle of thirty years' may not survive
the discovery that the statement which follows, although
presented as a personal reflection, is a commonplace of his age.

Browne's habit of assuming proprietary rights over ideas
which formed part of the intellectual furniture of his times
might be excused by the community of thought which he

enjoyed with his contemporaries; but his implicit assertion of ownership seems often to exclude any notion of sharing. 'I could never divide myself from any man upon the difference of an opinion', he begins one such passage, 'or be angry with his judgement for not agreeing with me in that from which, perhaps within a few days, I should dissent myself.' The note of emphasis which Browne strikes in his opening words invites us to suppose that he is putting forward a private opinion; and it may come as a surprise to learn that Montaigne had expressed the same idea in the *Apologie de Raimond Sebond*. Repeatedly Browne's manner leads us to suppose that he is voicing an eccentric personal philosophy when in fact he is repeating points of Scholastic doctrine or ideas caught up from the mass of meditative writing which he had obviously studied. Thus he remarks of angels, 'I believe they have an extemporary knowledge, and upon the first motion of their reason do what we cannot without study or deliberation; that they know things by their forms, and define by specifical differences what we describe by accidents and properties'. From its tone of weighty judgement, most readers would take the passage to represent Browne's private conclusions upon angelic nature. In fact, they are part of Thomist doctrine on the subject. The intellectual stock of the early seventeenth century was a common possession, but Browne assumes rights of enclosure which give him sole authority over the ideas which he adopts; tacitly claiming an originality of thought while often merely reviving controversial issues or familiar paradoxes. 'I cannot tell how to say that fire is the essence of hell' forces the reader to divide his attention between Browne and a problem in divinity which Browne adopts without acknowledgement to St Augustine. 'Nor truly can I peremptorily deny', another such passage begins, with an effect of ponderous deliberation, 'that the soul in this her sublunary estate is wholly and in all her acceptions inorganical.' None of the great Schoolmen of the thirteenth century could have denied it. Browne is asserting a commonplace; yet presenting the idea as the conclusion of a metaphysical debate within himself which has newly reached its terminus.

It would be unfair to suggest that Browne is deliberately encouraging his readers to accept a false estimate of himself as philosopher and eccentric. The sense of unique importance which his book communicates must have been real enough to him. 'I thank God, and with joy I mention it, I was never afraid of hell', is one of many remarks which reflect Browne's conviction that his private feelings and reflections have a crucial value. 'I cannot dream,' he begins another emphatic statement, 'there should be at the last day any such judicial proceeding or calling to the bar as indeed the Scripture seems to imply.' His incredulity, rather than the point of theological doctrine, is made the focus of attention; yet the larger claim he makes for himself towards the end of *Religio Medici*, perhaps the least expected of Browne's self-disclosures, allows us to suppose that he was innocent of arrogant intention. 'I thank God,' he begins in familiar style, 'amongst those millions of vices I do inherit and hold from Adam, I have escaped one, and that a mortal enemy to charity; the first and father sin, not only of man but of the devil, pride.' This unqualified confidence is hardly the idiom of humility. His satisfaction at having overcome insular prejudice towards foreign cooking—'I wonder not at the French for their dishes of frogs, snails and toadstools'—and at his command of six languages, 'besides the jargon and patois of several provinces', is attractive in its naïvety; but some of the larger claims which Browne makes for himself were understandably greeted with scepticism. 'I am in the dark to all the world, and my nearest friends behold me but in a cloud', is a muted version of the extravagant declaration which provoked Sir Kenelm Digby to mock at his pose of lofty inscrutability: 'Men that do look upon my outside, perusing only my condition and fortunes, do err in my altitude; for I am above Atlas' shoulders.'

If we are to understand *Religio Medici* we must accept the strain of complacency and self-absorption which runs through the book, recognising that its florid manner represents through literary gesture the philosophical attitude which Browne's ideas are intended to define. It is more difficult to be indulgent toward the inconsistency of thought which critical scrutiny brings to light. As we might suspect from the occasional incompatibility of the claims which Browne makes for his temperament, he can be moved to identify himself with different philosophical attitudes, apparently upon whimsical impulse; adopting as his own whatever odd and paradoxical notions do not actively conflict with his religious faith. The absence of an organised scheme of ideas in *Religio Medici*, the sudden broaching of new topics, and the digressions which interrupt its arguments, all suggest Browne's readiness to deflect the course of discussion upon whatever detached curiosities of thought his reflections can absorb. Rather than a *table raisonnée* of its author's beliefs, the book is a generally arbitrary collection, arranged in the haphazard order of rumination. It is part of the attraction of *Religio Medici* that its ideas are presented so informally, and part of its value that Browne takes into himself the intellectual plankton about him in such indiscriminate and impartial mouthfuls. As a representative cross-section of the ideas with which educated consciousness was stocked during the early seventeenth century, *Religio Medici* has no serious rival in our literature.

This usefulness is qualified by Browne's unresponsiveness to the new currents of thought which had been stirring throughout his lifetime. Not all critics share this view,[1] but Browne's fidelity to the Renaissance image of man is the index of his intellectual conservatism. In the year he was born, *The*

---

[1] See, for instance, Joan Bennett, *Sir Thomas Browne* (Cambridge, 1962).

Advancement of Learning was published; but Browne passes over Bacon and Harvey's discovery of the circulation of the blood with the same silence; and he alludes to the Copernican theory only to ridicule it. His indifference towards the new learning is implicit in his approving references to figures and propositions of the old philosophy which provide the substance of his discussions. It is well to be clear on this point. In recent years, with a reading public conditioned to think of the seventeenth century as the great transitional period between medieval and modern, Browne has been presented as a significant thinker, responsive to the intellectual cross-currents of an age whose divided sensibility is reflected in his work. This assumption seems immediately plausible if the famous passage about man 'the true amphibium' living simultaneously in 'divided and distinguished worlds' is taken as symbolising Browne's own position, midway between the opposing forces of Scholasticism and empirical science. But the passage does not invite this interpretation. What Browne himself intended by these phrases is shown a little earlier in Section 34, where he describes man more precisely, as 'that amphibious piece between a corporal and spiritual essence; that middle frame that links those two together'. This is a central commonplace of the old philosophy, and entirely characteristic of Browne's intellectual position. Nowhere does he betray any awareness of living in 'the divided and distinguished worlds' which modern criticism sees him straddling. Analysis of Browne's ideas reveals not merely a lack of sympathy with the emerging attitudes of the new philosophy, but no real appreciation of their existence. Had he possessed any talent for original metaphysical enquiry, his temperamental affinity with the old world of belief might have proved frustrating; but as Browne diverts himself with the oddities of Scholastic speculation he radiates intellectual contentedness: a mood whose untroubled confidence most readers find it hard to resist. If his faith had been recently shaken, it had re-established itself as a monumental assurance that was immune to the restless enquiries and discoveries of the new age. The picture of Browne as the 'great and true amphibium' of seventeenth-century outlook must be discarded as misleading and unfair to Browne, who makes no such claim for himself.

If *Religio Medici* can be considered as a work of popular philosophy, its valid interest lies elsewhere: in its review of traditional beliefs by an erudite but simple and uncritical mind, at a time when the authority of Scholasticism, immobilised by the obstinacy of its own tenets and unable to renew its intellectual vitality, was approaching collapse. *Religio Medici* ignores this steady decline, browsing at leisure through ideas drawn from Browne's reading of the ancient philosophers, the Christian Fathers, the Schoolmen, and Holy Writ. Some of these ideas retain the curious fascination which evidently attracted Browne; but we can discern, as he could not, their lack of underlying intention. Scholasticism was no longer presenting itself as a coherent and purposeful system of thought, but had degenerated into an undefined agglomeration of picturesque notions and fancies. As such, it was well suited to the indulgent and generally superficial form of enquiry which Browne undertakes in his book. His pronounced taste for curious, contradictory and insoluble points of doctrine might be taken as the mark of a questioning mind; but Browne is content to exhibit the anomalies of Scholastic belief, and does not intend to resolve the enigmatic issues which he raises. 'How America abounded with beasts of prey and noxious animals,' he muses, 'yet contained not in it that necessary creature a horse, is very strange.' He makes no further attempt to probe the mystery, and offers no hypothesis. A bare monosyllabic comment represents Browne's intellectual interest in a teasing point of natural history on which, until we know him better, we might have expected some considered judgement.

The impression of a mind disinclined to undertake any serious enquiry into the nature of the physical world is strengthened by a well-known passage of Browne's book. 'It is the microcosm of mine own frame that I cast mine eyes on', he admits without embarrassment; 'for the other, I use it but like my globe, and turn it round sometimes for my recreation.' The admission condemns him of intellectual dilettantism; a mere sporting with ideas which may produce paradoxes and enigmas, but—as Bacon had complained of Scholasticism —nothing of substance or profit. 'There are a bundle of curiosities,' Browne remarks after listing and answering a succession of fantastic propositions in divinity, 'proposed and discussed by men of the most supposed ability, which indeed are not worthy of our vacant hours, much less our more serious studies.' But the example he himself gives of 'most excellent speculation' in the causes, nature and influence of eclipses is, in point of any practical outcome, as worthless as the discussion whether Adam was a hermaphrodite. 'I could never content my contemplation,' he declares, 'with those general pieces of wonder, the flux and reflux of the sea, the increases of Nile, the conversion of the needle to the north.'

Such contemplation undertakes no practical end, but the deliberate mystifying of intelligence to the point where it surrenders and seeks relief in the O altitudo! of St Paul. Nothing could be more opposed to the spirit of empirical science than the positive pleasure which Browne derives from the incomprehensible and the insoluble. His taste for baffling enigma is reflected in the form of one of his most familiar rhetorical figures: the oblique question which tails off, sometimes lamely, into a shelving or evading of the issue. In a famous passage of Urn Burial, 'What song the Sirens sang, or what name Achilles assumed when he hid himself among women'—enquiries intellectually fruitless in themselves, however engagingly whimsical—leads to nothing more than an admission of difficulty: 'though puzzling questions,' and an unsubstantiated hint of eventual solution: 'are not beyond all conjecture'. The point of the sentence barely extends beyond Browne's statement of his odd conundrums. These are not always directly proposed. 'Whether Eve was framed out of the right side of Adam', is a proposition immediately cancelled by the statement of an enigma which obscures the puzzle: 'I dispute not, because I stand not yet assured which is the right side of man, nor whether there be any such distinction in nature.' Whether left and right have any significance 'in nature', that is, outside the scheme of human reference to which Browne has appealed, is a purely gratuitous point of discussion introduced only for its throw-away effect; a metaphysical red herring which allows him to substitute a distracting paradox for the direct answer which his first enquiry demands. 'That she was edified out of the rib of Adam, I believe', he continues, completing the rhetorical design of his sentence, 'yet raise no question who shall arise with that rib at the Resurrection.' This statement is true only in the sense that Browne does not actively debate the enigmatic issue which he brings before us, but again deflects attention from one curious problem by introducing another, to be left unattempted like the first. What to an uncritical reading might appear a process of connected thought is revealed as a casual succession of ideas linked only by a common quality of oddness, and leading nowhere. Browne is attempting to prove nothing but an intellectual sympathy with the mental world he inhabits; a storehouse of queer and baffling ideas whose oddity is their entire *raison d'être*.

His attachment to medieval thought is not simply a matter of his respect for beliefs and authorities which had already come under attack from the empiricists, but of the mental habits which direct his thinking. Thirty years after Bacon had criticised the Schoolmen's interest in final causes for its interception of 'enquiry of all real and physical causes', Browne is able to affirm without embarrassment, 'This is the cause I grope after in the works of nature'. Bacon, it is true, had allowed that theologians and metaphysicians might pursue this form of enquiry, but Browne is unable to detach his scientific thinking from this medieval tradition. His belief that man had been created to render homage to his Maker, and that this duty was properly answered in the astonished wonderment that must follow 'judicious enquiry into his acts', characterises the contemplative method of Scholastic philosophy which Bacon had berated for its fruitlessness. Browne's turning of the same gratified astonishment upon himself, 'that bold and adventurous piece of nature', and the habitual acquiescence in mystery which marks his outlook, declare an affinity with the old philosophy that goes much deeper than simple agreement with its postulates.

His speculation over the fire of hell in Section 50 of *Religio Medici* shows how closely his thinking is directed by the inductive methods of Scholasticism. Thus he debates whether a material fire is to be expected, and by what means the insubstantial soul might be consumed, with no acknowledgement that his conclusions are as hypothetical as the basis of his argument is suppositious. A modern scientist might as well try to determine by disputation whether the inhabitants of Venus use coat-hangers. It is a good example of the disease of Scholastic learning which Bacon attributes to 'a kind of adoration of the mind and understanding of man': the error of the intellectualists who 'by continual meditation and agitation of wit do urge and as it were invoke their own spirits to divine and give oracles unto them, whereby they are deservedly deluded'.[1]

Reason, as scientific thought was to understand the faculty, is not exalted but depreciated by Browne in his eagerness to embrace the enigmas which form the staple of his intellectual

[1] *Advancement of Learning*, I, v, 7.

diet, and to prove the impermeability of his religious faith. 'Since I was of understanding to know we know nothing,' he remarks with apparent satisfaction, 'my reason hath been more pliable to the will of faith.' At the end of a catalogue of discrepancies between Holy Writ and common experience he reaffirms this principle of belief: 'Yet do I believe that all this is true, which indeed my reason would persuade me to be false; and this I think is no vulgar part of faith, to believe a thing not only above but contrary to reason, and against the argument of our proper senses.'[1] Such an attitude might be commendable as an expression of simple piety; but Browne's determination to ignore his rational objections may also be read as a form of spiritual arrogance: a subordinating of reason not by the inner conviction of faith but by some lower faculty of the mind which rejects its authority as though on principle. When his reason offers him a scientific explanation of miracles and puzzles, as it does by suggesting that Elijah had ignited the sacrifice with the help of naphtha, Browne denounces it as the voice of Satan seeking to undermine his faith. When it offers to rebel in incredulity, he silences it with Tertullian's axiom, *Certum est, quia impossibile est*. When it confesses itself baffled, he rubs in the lesson as though he were admonishing some alien part of himself whose misdemeanours he was obliged to suffer: 'acquainting...reason how unable it is to display the visible and obvious effects of nature'. Nothing could show more clearly Browne's remoteness from the central premise of seventeenth-century science, that the laws governing natural processes and events could be determined by rational deduction from observation and experiment. By harassing and immobilising his reason, Browne's faith preserves the superstitious obscurity of mind which his intellectual comfort demands.

---

[1] It is interesting to note Browne's inconsistency in his borrowings from Scholasticism. His particular tenets are debased Thomism; but this general position forms part of Ockhamist theory.

Perhaps this is why the conclusion of his arguments is habitually the statement of a paradox, and not the elucidation of the problem he has raised. 'And so nothing becomes something, and omneity informed nullity into an essence', he winds up Section 35; and his reflections upon identity and uniqueness are concluded in a preposterous sophism: 'And thus is man like God, for in the same things that we resemble him we are utterly different from him.' We learn to recognise the preliminary sign of an incipient paradox: 'In brief, we have devoured ourselves, and yet do live and remain ourselves'; 'In brief, we are all monsters'; 'In brief, conceive light invisible, and that is a spirit'. This trick of style, supplied automatically to accompany Browne's metaphysical sleight-of-hand, suggests both the rigidity and the shallowness of his mental habit. How little controlled purpose directs his argument he shows in his discussion of general beauty, interpolated into Section 16, where his unwearied searching for paradox leads him round in a circle of contradiction: 'There is therefore no deformity but in monstrosity, wherein notwithstanding there is a kind of beauty.' Another cautionary signal introduces a still more ambitious exercise in paradox: 'To speak yet more narrowly, there was never anything ugly or unshapen but the Chaos, wherein notwithstanding, to speak strictly, there was no deformity because no form.' Despite the gravity which Browne confers upon his statement by claiming to weigh his words, this is hardly more than the answer to a riddle. In fact he is incapable of speaking strictly because, as Hobbes would have pointed out, he had neglected to define his terms. If there was no deformity in Chaos 'because no form', then Chaos was neither ugly nor misshapen, and Browne's demonstration falls to the ground. Nonetheless, in the previous paradox, deformity can be allowed to possess 'a kind of beauty'; here because Browne has deliberately confused the sense of words, making the concept 'monstrous' include 'beautiful' as an implicit idea: a piece of verbal juggling which has no reference to the reality denoted by these terms. In this respect Browne again follows debased Scholastic tradition; arguing from a tacit assumption that undefined terms could be treated as though they possessed the qualities of a corresponding part of reality, and that a demonstration by means of syllogism had all the force of experimental proof. Browne's thought is at the mercy of his vocabulary.

His fondness for incomprehensible ideas and his readiness to fall into an elevation of spirit are not to be mistaken for a strain of mysticism. In fact, the cautions which he imposes upon himself in the matter of religious controversy throw some doubt upon the stability of his faith. He thinks it wisdom to decline such disputes, he admits, 'especially upon a disadvantage': which seems to acknowledge the force of rational argument to unsettle his belief. Again, his contention that it is better to dispute with inferior minds, 'that the frequent spoils and victories over their reasons may settle in ourselves an esteem and confirmed opinion of our own', shows Browne both reaching after the assurance of personal superiority and guarding his private belief against the kind of assault which could unseat it. 'A man may be in as just possession of truth as of a city', he maintains, 'and yet be forced to surrender'; and without allowing himself time to scrutinise the contradiction which his thesis contains he goes on: 'If, therefore, there rise any doubts in my way, I do forget them; or at least defer them till my better settled judgement and more manly reason be able to resolve them.' Here reason is briefly allowed the important function of supporting faith. Browne is attempting to compensate for a lack of inward conviction by arguing himself into a complacency of belief. If the doubts that arise could undermine his faith despite his conviction of its truth, they may be ignored; and although Browne does not

acknowledge this, his surrender would then be an admission
that the weight of evidence and probability lay with the other
side. This doubtfulness of attitude, and his reluctance to sub-
mit his beliefs to the scrutiny of logic and reason, are more
than enough to discredit any suggestion of mysticism in
Browne's religious outlook. He appeals to the incomprehensible and the inexplicable as a refuge from the rational
analysis that could bring his intellectual security into peril.
''Tis my solitary recreation,' he tells us, 'to pose my apprehension with those involved enigmas and riddles of the Trinity,
with Incarnation and Resurrection.' How literal was his
approach to one of the central mysteries of his faith we see
from Browne's ruminations on the Trinity in Section 12,
where he remarks on the oddity that 'though in a relative way
of Father and Son, we must deny a priority'. The Incarnation
and the Resurrection are treated as advanced conundrums
whose very nature prohibits their being unpicked by human
understanding. The possibility of mystical enlightenment
seems not to occur to Browne, whose only approach to these
mysteries is one of an intellectual curiosity which, knowing
itself defeated from the outset, contents itself with a display
of uncommitted speculation.

What must finally dispose of the idea that Browne was
intellectually sympathetic toward the new science is his
tendency to abandon natural philosophy for divinity when it
comes under stress. His taste for mystery, his active preference
for the incomprehensible, reflect more than a childlike delight
in oddity and paradox. Science offers him no positive assurance, no security of belief; whereas religious orthodoxy
provides a protecting wall of dogma which Browne adopts
with fixed determination. 'No man shall reach my faith unto
another article, or command my obedience to a canon more',
he affirms of his Anglican orthodoxy; and again, 'Where the
Scripture is silent, the Church is my text; where that speaks,

'tis but my comment'. The liberty of speculation which he allows himself beyond this double authority, 'to play and expatiate' with capricious notions, will be withdrawn immediately he recognises the taint of heresy in his enquiries. The appearance of vigorous, even daring intellectual activity which he contrives is thus heavily compromised by his determination to remain inside the pale of strict conformity. The eccentric and paradoxical ideas which he submits to his readers are, by this admission, concerned only with peripheral issues of religious faith. The solid ground of orthodoxy remains unchallenged.

Browne's scientific outlook remains subservient to this absolute authority. He shows no inclination to protest against the subordinate position which his own studies, anatomy and medicine, were obliged to accept, but seconds the attitude of the Galenists who were led from their investigation of man's intricate bodily structure to admiration of the creating hand which had shaped him. More than this, Browne habitually retreats from scientific observation to religious awe; a shift of purpose which burks the issue confronting him as a potential man of science. His remark about the 'strange and mystical transmigrations' of silkworms which turned his philosophy into divinity is characteristic of Browne's lack of curiosity about the nature of the physical change involved. Their metamorphosis interests him only as an emblem or 'mystical type' of the greater change which man must undergo at death. In at least ten passages of *Religio Medici* that associate natural philosophy and divinity, Browne leaves no doubt of his respect for the authority of the senior subject. 'In philosophy, where truth seems double-faced,' he asserts, 'there is no man more paradoxical than myself; but in divinity I love to keep the road, and...follow the great wheel of the Church, by which I move; not reserving any proper poles or motion from the epicycle of my own brain. By this means I leave no gap for heresies, schisms or errors.' Philosophy and science may permit of playful speculation, but divinity demands strict and serious conformity. For this reason Browne is prepared to credit what he knows to be impossible as a demonstration of his faith, illogically preferring belief when it is not supported by material evidence. The testimony of the senses, so important to empirical science, he slights as a factor of little account: ''Tis an easy and necessary belief to credit what our eye and sense have examined': and without showing any propensity for analysing such evidence. When he adopts the manner of a logical argument, he usually does little more than repeat his thesis in the form of a demonstration; asserting, in effect, that occurrences take place because they happen. This forms the substance of his remark about the observed behaviour of the sun in Section 16: 'To make a revolution every day is the nature of the sun, because it is that necessary course which God hath ordained it.' Browne's 'because', acting as the rhetorical hinge of the statement, gives the impression that he is offering a scientific explanation where in fact he is merely varying the terms in which the phenomenon is described. The same iteration of idea, varied but not developed, and a similar restatement of premise as conclusion to his argument, recur in the more extended passage on vicissitude in Section 17:

All cannot be happy at once; for, because the glory of one state depends upon the ruin of another, there is a revolution and vicissitude of their greatness, which must obey the swing of that wheel not moved by Intelligences but by the hand of God, whereby all estates arise to their zenith and vertical points, according to their predestinated period. For the lives not only of men but of commonweals and the whole world, run not upon an helix that still enlargeth, but on a circle; where arriving to their meridian they decline in obscurity, and fall under the horizon again.

What follows the initial thesis, 'all cannot be happy at once', we may mistake for the first phase of a demonstration; 'for, because the glory of one state depends upon the ruin of another'. In fact this is a new proposition; and without attempting to prove either, Browne proceeds to develop the second idea rhetorically through the figure of a turning wheel. Then follows a reaffirmation of the original thesis in different terms: 'the lives not only of men but of commonweals, and the whole world, run...on a circle'. This proposition too he demonstrates not by logical argument but rhetorically, making the movement of his sentence match the sense it conveys. Browne has said very little more than is contained in the six opening words of the paragraph. The rest amounts to a free variation on this theme, governed by the still unsupported assertion that 'the glory of one state depends upon the ruin of another'. He has already assumed the truth of the hypothetical law he should be proving, and the terms in which he describes its working can only seem to confirm its existence. Intellectually, nothing is happening: the process of thought is lifeless. Browne's impressive command of rhetoric, and especially his feeling for what Thomas Sprat was to call a glorious pomp of words, helps to conceal the absence of creative intellectual activity in his writing, or to compensate for its absence. On this showing it is difficult to believe that Browne would have been capable of making any useful contribution to the thought of the new age, even had he been sympathetic towards its aims.

The rhetorical power of his writing accounts most readily for Browne's general popularity, and for the respect which he has attracted as a major figure of the great age of English prose. The impression of ponderous reflection which he creates by the uninterrupted movement of long sentences and weighty vocabulary may be untrustworthy, but there is no denying the immediate effect of majestic utterance:

Now, for that immaterial world, methinks we need not wander so far as beyond the first moveable; for even in this material fabric the spirits walk as freely exempt from the affections of time, place and motion as beyond the extremest circumference.

This says little more than that spirits enjoy the same freedom within the material world as without; but even if we feel that Browne's grandiloquence does not justify itself we may find it hard to resist being carried forward by the surge of rhetoric which determines the shape of his sentence. To an uncritical appreciation it does not matter that the intellectual content of such a passage is nugatory, or that Browne's majestic terms cover nothing more than a familiar commonplace. The language and movement of his sentences communicate an experience that does not call for intellectual stiffening:

Now, for those walls of flesh wherein the soul doth seem to be immured before the Resurrection, it is nothing but an elemental composition, and a fabric that must fall to ashes.

The scriptural commonplace, 'All flesh is grass', puts the point more succinctly, but without the overtones of authority and finality which Browne confers upon it. His transformation of this simple statement into a massively emphatic commentary is typical of the treatment which the philosophical ideas of *Religio Medici* are made to undergo. The statement 'all flesh is grass' is presented in terms of an occurrence; for it is an effect of Browne's rhetorical expansion to present the original idea as though he were describing a metaphysical event. In the passage last quoted he brings about this impression by making his sentence climb arduously to the end of the opening clause, from which it freewheels downwards as though by its own heavy momentum. The simple content of the sentence shows that the rhetorical figure is irrelevant. Its heavy rising and falling corresponds to nothing in the idea which it expresses; but the gravity which the statement acquires may easily persuade us that we are participating in an exercise of thought without the fatigue of ratiocination.

His mind, Browne declares modestly, is a treasure-house of knowledge, and his studies not intended merely for his private benefit, but for those 'that study not for themselves'. In *Religio Medici* he spends this accumulated learning generously, though without being able to separate his knowledge from the intrusive self who has acquired it. His peronality acts as an adhesive force holding together the curious mass of ideas of which his book is composed. *Religio Medici* was written when Scholasticism was becoming inert: a crumbling system of thought from which Browne could pilfer metaphysical odds and ends to stock a private collection of outlandish notions, but incapable of providing the nucleus of an independent philosophy. The ideas of *Religio Medici* are important chiefly as they reveal the idiosyncrasies of its author's outlook; marking out the intellectual scope of his mind without evolving a coherent attitude to life. The post-Renaissance image of man which emerges is synthetic; a compilation of attributes, sometimes contradictory, which does not carry the impress of imaginative conviction. With this awkwardly histrionic figure Browne identifies himself, supplying its huge gestures and cosmic significance from an exhausted tradition; yet managing to infuse his own vitality into the moribund oddities of thought which he brings together as its context. This is his principal achievement: not to have promoted a happy alliance between the old philosophy and the revolutionary ideas of his age, but to have reinvigorated some of the attitudes of late Scholasticism to which his innate conservatism and love of mystery committed him. Three centuries of popularity have confirmed the interest of his ruminations; not for any philosophical depth which Browne may have supposed them to possess, but for their reflection of a mind at peace within its own borders, fascinated by the intricacies of a system of ideas without influence upon the creative purposes of the age.

JAMES WINNY

RELIGIO MEDICI

TO THE READER

Certainly that man were greedy of life who should desire to live when all the world were at an end, and he must needs be very impatient who would repine at death in the society of all things that suffer under it. Had not almost every man suffered by the press, or were not the tyranny thereof become universal, I had not wanted reason for complaint; but in times wherein I have lived to behold the highest perversion of that excellent invention, the name of his Majesty defamed, the honour of Parliament depraved, the writings of both depravedly, anticipatively, counterfeitly imprinted, complaints may seem ridiculous in private persons, and men of my condition may be 10 as incapable of affronts as hopeless of their reparations. And truly, had not the duty I owe unto the importunity of friends, and the allegiance I must ever acknowledge unto truth prevailed with me, the inactivity of my disposition might have made these sufferings continual; and time, that brings other things to light, should have satisfied me in the remedy of its oblivion. But because things evidently false are not only printed, but many things of truth most falsely set forth, in this latter I could not but think myself engaged: for though we have no power to redress the former, yet in the other the reparation being within ourselves, I have at present represented unto the world a full 20 and intended copy of that piece which was most imperfectly and surreptitiously published before.

This, I confess, about seven years past, with some others of affinity thereto, for my private exercise and satisfaction, I had at leisurable hours composed; which being communicated unto one, it became common unto many, and was by transcription successively corrupted, until it arrived in a most depraved copy at the press. He that shall peruse that work, and shall take notice of sundry particularities and personal expressions therein, will easily discern the intention was not public; and being a private exercise directed to myself, what is 30 delivered therein was rather a memorial unto me than an example or rule unto any other. And therefore if there be any singularity therein correspondent unto the private conceptions of any man, it doth not

I

advantage them; or if dissentaneous thereunto, it no way overthrows them. It was penned in such a place, and with such disadvantage, that—I protest—from the first setting of pen unto paper I had not the assistance of any good book whereby to promote my invention or relieve my memory; and therefore there might be many real lapses therein, which others might take notice of, and more than I suspected myself. It was set down many years past, and was the sense of my conceptions at that time, not an immutable law unto my advancing judgement at all times; and therefore there might be many things 10 therein plausible unto my past apprehension which are not agreeable unto my present self. There are many things delivered rhetorically, many expressions therein merely tropical, and as they best illustrate my intention; and therefore also there are many things to be taken in a soft and flexible sense, and not to be called unto the rigid test of reason. Lastly, all that is contained therein is in submission unto maturer discernments; and, as I have declared, shall no further father them than the best and learned judgements shall authorise them; under favour of which considerations I have made its secrecy public, and committed the truth thereof to every ingenuous reader.

THOMAS BROWNE

RELIGIO MEDICI

THE FIRST PART

Section 1. For my religion, though there be several circumstances that might persuade the world I have none at all—as the general scandal of my profession, the natural course of my studies, the indifferency of my behaviour and discourse in matters of religion, neither violently defending one, nor with that common ardour and contention opposing another—yet in despite hereof I dare, without usurpation, assume the honourable style of a Christian. Not that I merely owe this title to the font, my education, or clime wherein I was born, as being bred up either to confirm those principles my parents instilled into my unwary understanding, or by a general consent proceed in the religion of my country; but that having in my riper years and confirmed judgement seen and examined all, I find myself obliged by the principles of grace and the law of mine own reason to embrace no other name but this. Neither doth herein my zeal so far make me forget the general charity I owe unto humanity, as rather to hate than pity Turks, infidels, and—what is worse—the Jews; rather contenting myself to enjoy that happy style than maligning those who refuse so glorious a title.

Section 2. But because the name of a Christian is become too general to express our faith—there being a geography of religions as well as lands, and every clime distinguished not only by their laws and limits, but circumscribed by their doctrines and rules of faith—to be particular, I am of that reformed new-cast religion, wherein I mislike nothing but the name; of the same belief our Saviour taught, the Apostles disseminated, the Fathers authorised, and the martyrs con-

3

firmed; but by the sinister ends of princes, the ambition and avarice of prelates, and the fatal corruption of times, so decayed, impaired and fallen from its native beauty that it required the careful and charitable hands of these times to restore it to its primitive integrity. Now the accidental occasion whereon, the slender means whereby, the low and abject condition of the person by whom so good a work was set on foot, which in our adversaries beget contempt and scorn, fills me with wonder, and is the very same objection the insolent
10 pagans first cast at Christ and his disciples.

Section 3. Yet have I not so shaken hands with those desperate resolutions—who had rather venture at large their decayed bottom than bring her in to be new trimmed in the dock; who had rather promiscuously retain all than abridge any, and obstinately be what they are than what they have been—as to stand in diameter and sword's point with them. We have reformed from them, not against them; for, omitting those improperations and terms of scurrility betwixt us, which only difference our affections and not our cause, there
20 is between us one common name and appellation, one faith and necessary body of principles common to us both; and therefore I am not scrupulous to converse and live with them, to enter their churches in defect of ours, and either pray with them or for them. I could never perceive any rational consequence from those many texts which prohibit the children of Israel to pollute themselves with the temples of the heathens; we being all Christians, and not divided by such detested impieties as might profane our prayers, or the place wherein we make them; or that a resolved conscience may not adore
30 her Maker anywhere, especially in places devoted to his service; where, if their devotions offend him, mine may please him, if theirs profane it, mine may hallow it. Holy water and crucifix—dangerous to the common people—deceive not my

4

judgement nor abuse my devotion at all. I am, I confess, naturally inclined to that which misguided zeal terms superstition. My common conversation I do acknowledge austere, my behaviour full of rigour, sometimes not without morosity; yet at my devotion I love to use the civility of my knee, my hat and hand, with all those outward and sensible motions which may express or promote my invisible devotion. I should violate my own arm rather than a church, nor willingly deface the memory of saint or martyr. At the sight of a cross or crucifix I can dispense with my hat, but scarce with the 10 thought and memory of my Saviour. I cannot laugh at, but rather pity the fruitless journeys of pilgrims, or contemn the miserable condition of friars; for though misplaced in circumstance, there is somewhat in it of devotion. I could never hear the Ave Marie bell without an elevation, or think it a sufficient warrant, because they erred in one circumstance, for me to err in all—that is, in silence and dumb contempt. Whilst therefore they directed their devotions to her, I offered mine to God, and rectified the errors of their prayers by rightly ordering mine own. At a solemn procession I have wept abundantly 20 while my consorts, blind with opposition and prejudice, have fallen into an excess of scorn and laughter. There are, questionless, both in Greek, Roman and African churches, solemnities and ceremonies whereof the wiser zeals do make a Christian use; and stand condemned by us, not as evil in themselves, but as allurements and baits of superstition to those vulgar heads that look asquint on the face of truth, and those unstable judgements that cannot consist in the narrow point and centre of virtue without a reel or stagger to the circumference. 30

Section 4. As there were many reformers, so likewise many reformations; every country proceeding in a peculiar method, according as their national interest together with their constitution and clime inclined them: some angrily and with extremity, others calmly and with mediocrity; not rending but easily dividing the community, and leaving an honest possibility of a reconciliation; which though peaceable spirits do desire, and may conceive that revolution of time and the mercies of God may effect, yet that judgement that shall consider the present antipathies between the two extremes, their contrarieties in condition, affection and opinion, may with the same hopes expect an union in the poles of heaven.

10 *Section 5.* But—to difference myself nearer, and draw into a lesser circle—there is no church wherein every point so squares unto my conscience, whose articles, constitutions and customs seem so consonant unto reason, and as it were framed to my particular devotion, as this whereof I hold my belief; the Church of England, to whose faith I am a sworn subject, and therefore in a double obligation subscribe unto her articles, and endeavour to observe her constitutions. No man shall reach my faith unto another article, or command my obedience to a canon more: whatsoever is beyond, as points in-
20 different, I observe according to the rules of my private reason or the humour and fashion of my devotion; neither believing this because Luther affirmed it, or disapproving that because Calvin hath disavouched it. I condemn not all things in the Council of Trent, nor approve all in the Synod of Dort. In brief, where the Scripture is silent, the Church is my text; where that speaks, 'tis but my comment; where there is a joint silence of both, I borrow not the rules of my religion from Rome or Geneva, but the dictates of my own reason. It is an unjust scandal of our adversaries and a gross error in ourselves
30 to compute the nativity of our religion from Henry VIII, who, though he rejected the Pope, refused not the faith of Rome; and effected no more but what his own predecessors desired and assayed in ages past, and was conceived the State of Venice

6

would have attempted in our days. It is as uncharitable a point in us to fall upon those popular scurrilities and opprobrious scoffs of the Bishop of Rome, whom as a temporal prince we owe the duty of good language. I confess there is cause of passion between us. By his sentence I stand excommunicate, and my posterity: 'heretic' is the best language he affords me, yet can no ear witness I ever returned to him the name of Antichrist, Man of Sin, or Whore of Babylon. It is the method of charity to suffer without reaction: those usual satires and invectives of the pulpit may by chance produce a good effect 10 on the vulgar, whose ears are opener to rhetoric than logic; yet do they in no wise confirm the faith of wiser believers, who know that a good cause needs not to be patroned by a passion, but can sustain itself upon a temperate dispute.

Section 6. I could never divide myself from any man upon the difference of an opinion, or be angry with his judgement for not agreeing with me in that from which, perhaps within a few days, I should dissent myself. I have no genius to disputes in religion, and have often thought it wisdom to decline them; especially upon a disadvantage, or when the cause of 20 truth might suffer in the weakness of my patronage. Where we desire to be informed 'tis good to contest with men above ourselves; but to confirm and establish our opinions 'tis best to argue with judgements below our own, that the frequent spoils and victories over their reasons may settle in ourselves an esteem and confirmed opinion of our own. Every man is not a proper champion for truth, nor fit to take up the gauntlet in the cause of verity. Many, from the ignorance of these maxims and an inconsiderate zeal unto truth, have too rashly charged the troops of error, and remain as trophies unto 30 the enemies of truth. A man may be in as just possession of truth as of a city, and yet be forced to surrender. 'Tis therefore far better to enjoy her with peace than to hazard her on a

7

battle. If, therefore, there rise any doubts in my way, I do forget them, or at least defer them till my better settled judgement and more manly reason be able to resolve them; for I perceive every man's own reason is his best Oedipus, and will, upon a reasonable truce, find a way to loose those bonds wherewith the subtleties of error have enchained our more flexible and tender judgements. In philosophy, where truth seems double-faced there is no man more paradoxical than myself; but in divinity I love to keep the road, and—though not in an implicit yet an humble faith—follow the great wheel of the Church, by which I move, not reserving any proper poles or motion from the epicycle of my own brain. By this means I leave no gap for heresies, schisms or errors, of which at present I hope I shall not injure truth to say I have no taint or tincture. I must confess my greener studies have been polluted with two or three—not any begotten in the latter centuries, but old and obsolete; such as could never have been revived but by such extravagant and irregular heads as mine; for indeed, heresies perish not with their authors, but like the river Arethusa, though they lose their currents in one place, they rise up again in another. One general council is not able to extirpate one singular heresy. It may be cancelled for the present, but revolution of time and the like aspects from heaven will restore it, when it will flourish till it be condemned again; for as though there were a metempsychosis, and the soul of one man passed into another, opinions do find, after certain revolutions, men and minds like those that first begat them. To see ourselves again we need not look for Plato's year. Every man is not only himself; there have been many Diogenes and as many Timons, though but few of that name. Men are lived over again; the world is now as it was in ages past. There was none then but there hath been someone since that parallels him, and is, as it were, his revived self.

8

Section 7. Now the first of mine was that of the Arabians, that the souls of men perished with their bodies, but should yet be raised again at the last day. Not that I did absolutely conceive a mortality of the soul; but if that were which faith, not philosophy, hath yet throughly disproved, and that both entered the grave together, yet I held the same conceit thereof that we all do of the body, that it should rise again. Surely it is but the merits of our unworthy natures if we sleep in darkness until the last alarum? A serious reflex upon my own unworthiness did make me backward from challenging this 10 prerogative of my soul. So I might enjoy my Saviour at last, I could with patience be nothing almost unto eternity. The second was that of Origen, that God would not persist in his vengeance for ever, but after a definite time of his wrath he would release the damned souls from torture. Which error I fell into upon a serious contemplation of the great attribute of God, his mercy; and did a little cherish it in myself because I found therein no malice, and a ready weight to sway me from that other extreme of despair, whereunto melancholy and contemplative natures are too easily disposed. A third 20 there is which I did never positively maintain or practise, but have often wished it had been consonant to truth and not offensive to my religion, and that is, the prayer for the dead; whereunto I was inclined from some charitable inducements, whereby I could scarce contain my prayers for a friend at the ringing out of a bell, or behold his corpse without an orison for his soul. 'Twas a good way, methought, to be remembered by posterity, and far more noble than an history. These opinions I never maintained with pertinacity, or endeavoured to inveigle any man's belief unto mine, nor so much as ever 30 revealed or disputed them with my dearest friends; by which means I neither propagated them in others nor confirmed them in myself; but suffering them to flame upon their own substance without addition of new fuel, they went out

9

insensibly of themselves. Therefore these opinions, though condemned by lawful councils, were not heresies in me but bare errors and single lapses of my understanding, without a joint depravity of my will. Those have not only depraved understandings but diseased affections which cannot enjoy a singularity without a heresy, or be the author of an opinion without they be of a sect also. This was the villainy of the first schism of Lucifer, who was not content to err alone but drew into his faction many legions of spirits; and upon this experience he
10 tempted only Eve, as well understanding the communicable nature of sin, and that to deceive but one was tacitly and upon consequence to delude them both.

Section 8. That heresies should arise we have the prophecy of Christ, but that old ones should be abolished we hold no prediction. That there must be heresies is true not only in our church, but also in any other. Even in doctrines heretical there will be super-heresies; and Arians not only divided from their church but also among themselves; for heads that are disposed unto schism, and complexionally propense to innovation, are
20 naturally indisposed for a community, nor will ever be confined unto the order or economy of one body. And therefore, when they separate from others they knit but loosely among themselves; nor contented with a general breach or dichotomy with their church, do subdivide and mince themselves almost into atoms. 'Tis true that men of singular parts and humours have not been free from singular opinions and conceits in all ages; retaining something not only beside the opinion of his own church or any other, but also any particular author; which notwithstanding, a sober judgement may do without offence or
30 heresy. For there is yet, after all the decrees of councils and the niceties of the Schools, many things untouched, unimagined, wherein the liberty of an honest reason may play and expatiate with security, and far without the circle of an heresy.

Section 9. As for those wingy mysteries in divinity and airy subtleties of religion, which have unhinged the brains of better heads, they never stretch the pia mater of mine. Methinks there be not impossibilities enough in religion for an active faith. The deepest mysteries ours contains have not only been illustrated but maintained by syllogism and the rule of reason. I love to lose myself in a mystery, to pursue my reason to an O *altitudo!* 'Tis my solitary recreation to pose my apprehension with those involved enigmas and riddles of the Trinity, with Incarnation and Resurrection. I can answer all 10 the objections of Satan and my rebellious reason with that odd resolution I learned of Tertullian, *Certum est, quia impossibile est.* I desire to exercise my faith in the difficultest points, for to credit ordinary and visible objects is not faith but persuasion. Some believe the better for seeing Christ's sepulchre, and when they have seen the Red Sea, doubt not of the miracle. Now contrary, I bless myself and am thankful that I lived not in the days of miracles, that I never saw Christ nor his disciples. I would not have been one of the Israelites that passed the Red Sea, nor one of Christ's patients on whom he wrought his 20 wonders. Then had my faith been thrust upon me, nor should I enjoy that greater blessing pronounced to all that believe and saw not. 'Tis an easy and necessary belief to credit what our eye and sense hath examined. I believe he was dead, buried, and rose again; and desire to see him in his glory rather than to contemplate him in his cenotaph or sepulchre. Nor is this much to believe. As we have reason, we owe this faith unto history: they only had the advantage of a bold and noble faith who lived before his coming, who upon obscure prophecies and mystical types could raise a belief and expect apparent 30 impossibilities.

Section 10. 'Tis true there is an edge in all firm belief, and with an easy metaphor we may say the sword of faith; but

in these obscurities I rather use it in the adjunct the Apostle gives it, a buckler; under which I perceive a wary combatant may lie invulnerable. Since I was of understanding to know we know nothing, my reason hath been more pliable to the will of faith. I am now content to understand a mystery without a rigid definition, in an easy and platonic description. That allegorical description of Hermes pleaseth me beyond all the metaphysical definitions of divines. Where I cannot satisfy my reason, I love to humour my fancy. I had as lief
10 you tell me that *anima est angelus hominis, est corpus Dei*, as *entelechia*; *lux est umbra Dei*, as *actus perspicui*. Where there is an obscurity too deep for our reason, 'tis good to sit down with a description, periphrasis, or adumbration; for by acquainting our reason how unable it is to display the visible and obvious effects of nature, it becomes more humble and submissive unto the subtleties of faith; and thus I teach my haggard and unreclaimed reason to stoop unto the lure of faith. I do believe there was already a tree whose fruit our unhappy parents tasted, though, in the same chapter where
20 God forbids it, 'tis positively said the plants of the field were not yet grown, for God had not caused it to rain upon the earth. I believe that the Serpent—if we shall literally understand it—from his proper form and figure, made his motion on his belly before the curse. I find the trial of the pucellage and virginity of women which God ordained the Jews is very fallible. Experience and history informs me that not only many particular women but likewise whole nations have escaped that curse of childbed which God seems to pronounce upon the whole sex. Yet do I believe that all this is true, which
30 indeed my reason would persuade me to be false; and this I think is no vulgar part of faith, to believe a thing not only above but contrary to reason, and against the arguments of our proper senses.

Section 11. In my solitary and retired imagination—*Neque enim cum porticus aut me lectulus accipit, desum mihi*—I remember I am not alone, and therefore forget not to contemplate him and his attributes who is ever with me; especially those two mighty ones, his wisdom and eternity. With the one I recreate, with the other I confound my understanding; for who can speak of eternity without a solecism, or think thereof without an ecstasy? Time we may comprehend; 'tis but five days older than ourselves, and hath the same horoscope with the world; but to retire so far back as to apprehend a beginning, to give such an infinite start forward as to conceive an end in an essence that we affirm hath neither the one nor the other, it puts my reason to St Paul's sanctuary, *O altitudo!* My philosophy dares not say the angels can do it. God hath not made a creature that can comprehend him: 'tis the privilege of his own nature. *I am that I am* was his own definition unto Moses; and 'twas a short one, to confound mortality that durst question God or ask him what he was. Indeed, he only is, all others have been and shall be; but in eternity there is no distinction of tenses. And therefore that terrible term predestination, which hath troubled so many weak heads to conceive, and the wisest to explain, is in respect of God no previous determination of our estates to come, but a definitive blast of his will already fulfilled, and at the instant that he first decreed it. For to his eternity, which is indivisible and altogether, the last trump is already sounded, the reprobates in the flame, and the blessed in Abraham's bosom. St Peter spake modestly when he said a thousand years to God are but as one day; for, to speak like a philosopher, those continued instants of time which flow into a thousand years make not to him one moment. What to us is to come, to his eternity is present; his whole duration being but one permanent point, without succession, parts, flux or division.

Section 12. There is no attribute that adds more difficulty to the mystery of the Trinity, where, though in a relative way of Father and Son, we must deny a priority. I wonder how Aristotle should conceive the world eternal, or how he could make good two eternities. His similitude of a triangle comprehended in a square doth somewhat illustrate the trinity of our souls, and that the triple unity of God: for there is in us not three but a trinity of souls; because there is within us, if not three distinct souls, yet different faculties that can and do subsist apart in different subjects, and yet in us are so united as to make but one soul and substance. If one soul were so perfect as to inform three distinct bodies, that were a petty trinity: conceive the distinct number of three, not divided nor separated by the intellect, but actually comprehended in its unity, and that is a perfect trinity. I have often admired the mystical way of Pythagoras, and the secret magic of numbers: 'beware of philosophy' is a precept not to be received in too large a sense; for in this mass of nature there is a set of things which carry in their front—though not in capital letters, yet in stenography and short characters—something of divinity, which to wiser reasons serve as luminaries in the abyss of knowledge, and to judicious beliefs as scales and roundles to mount the pinnacles and highest pieces of divinity. The severe Schools shall never laugh me out of the philosophy of Hermes, that this visible world is but a picture of the invisible, wherein, as in a portrait, things are not truly but in equivocal shapes, and as they counterfeit some more real substance in that invisible fabric.

Section 13. That other attribute wherewith I recreate my devotion is his wisdom, in which I am happy; and for the contemplation of this only do not repent me that I was bred in the way of study. The advantage I have of the vulgar, with the content and happiness I conceive therein, is an ample recompense for all my endeavours, in what part of knowledge soever. Wisdom is his most beauteous attribute; no man can attain unto it, yet Solomon pleased God when he desired it. He is wise because he knows all things; and he knoweth all things because he made them all; but his greatest knowledge is in comprehending that he made not; that is, himself. And this is also the greatest knowledge in man. For this do I honour my own profession, and embrace the counsel even of the devil himself. Had he read such a lecture in Paradise as he did at Delphos, we had better known ourselves; nor had we stood in fear to know him. I know that he is wise in all, wonderful in that we conceive, but far more in what we comprehend not; for we behold him but asquint, upon reflex or shadow. Our understanding is dimmer than Moses' eye; we are ignorant of the back-parts of God and the lower side of his divinity; therefore to pry into the maze of his counsels is not only folly in man, but presumption even in angels. There is no thread or line to guide us in that labyrinth. Like us, they are his servants, not his senators. He holds no council but that mystical one of the Trinity, wherein though there be three persons there is but one mind, that decrees without contradiction; nor needs he any. His actions are not begot with deliberation; his wisdom naturally knows what's best; his intellect stands ready fraught with the superlative and purest Ideas of goodness. Consultation and election, which are two motions in us, are but one in him, his actions springing from his power at the first touch of his will. These are contemplations metaphysical: my humble speculations have another method, and are content to trace and discover those impressions he hath left on his creatures, and the obvious effects of nature. There is no danger to propound these mysteries, no *sanctum sanctorum* in philosophy. The world was made to be inhabited by beasts, but studied and contemplated by man: 'tis the debt of our reason we owe unto God, and the homage we pay for not being beasts. Without this, the world is still as though it had not been, or as it was before the sixth day, when as yet there was not a creature that could conceive or say there was a world. The wisdom of God receives small honour from those vulgar heads that rudely stare about, and with a gross rusticity admire his works. Those highly magnify him whose judicious enquiry into his acts, and deliberate research of his creatures, return the duty of a devout and learned admiration. Therefore,

10   Search while thou wilt, and let thy reason go
      To ransom truth, even to the abyss below.
      Rally the scattered causes, and that line
      Which nature twists, be able to untwine.
      It is thy Maker's will, for unto none
      But unto reason can he e'er be known.
      The devils do know thee, but those damned meteors
      Build not thy glory, but confound thy creatures.
      Teach my endeavours so thy works to read
      That, learning them, in thee I may proceed.
20   Give thou my reason that instructive flight
      Whose weary wings may on thy hands still light.
      Teach me to soar aloft, yet ever so,
      When near the sun, to stoop again below.
      Thus shall my humble feathers safely hover,
      And, though near earth, more than the heavens discover.
      And then at last, when homeward I shall drive,
      Rich with the spoils of nature, to my hive,
      There will I sit like that industrious fly,
      Buzzing thy praises, which shall never die
30   Till death abrupts them, and succeeding glory
      Bid me go on in a more lasting story.

And this is almost all wherein an humble creature may endeavour to requite and someway to retribute unto his Creator: for if not he that saith, Lord, Lord, but he that doth the will of the Father shall be saved, certainly our wills must be our performances, and our intents make out our actions. Otherwise our pious labours shall find anxiety in their graves, and our best endeavours not hope but fear a resurrection.

Section 14. There is but one first cause and four second causes of all things. Some are without efficient, as God; others without matter, as angels; some without form, as the first matter; but every essence, created or uncreated, hath its final cause and some positive end, both of its essence and operation. This is the cause I grope after in the works of nature; on this 10 hangs the providence of God. To raise so beauteous a structure as the world and the creatures thereof was but his art; but their sundry and divided operations, with their predestinated ends, are from the treasury of his wisdom. In the causes, nature, and affections of the eclipse of the sun and moon there is most excellent speculation; but to propound farther, and to contemplate a reason why his providence hath so disposed and ordered their motions in that vast circle as to conjoin and obscure each other, is a sweeter piece of reason, and a diviner point of philosophy. Therefore sometimes, and in some 20 things, there appears to me as much divinity in Galen's books *De usu partium* as in Suarez' metaphysics. Had Aristotle been as curious in the enquiry of this cause as he was of the other, he had not left behind him an imperfect piece of philosophy, but an absolute tract of divinity.

Section 15. *Natura nihil agit frustra* is the only indisputable axiom in philosophy. There are no grotesques in nature, nor anything framed to fill up empty cantons and unnecessary spaces. In the most imperfect creatures—and such as were not preserved in the Ark, but having their seeds and principles in 30 the womb of nature, are everywhere where the power of the sun is—in these is the wisdom of his hand discovered. Out of

this rank Solomon chose the object of his admiration. Indeed, what reason may not go to school to the wisdom of bees, ants and spiders? What wise hand teacheth them to do what reason cannot teach us? Ruder heads stand amazed at those prodigious pieces of nature, whales, elephants, dromedaries and camels. These, I confess, are the colossus and majestic pieces of her hand; but in these narrow engines there is more curious mathematics, and the civility of these little citizens more neatly sets forth the wisdom of their Maker. Who admires not
10 Regio-Montanus' fly beyond his eagle, or wonders not more at the operation of two souls in those little bodies than but one in the trunk of a cedar? I could never content my contemplation with those general pieces of wonder, the flux and reflux of the sea, the increase of Nile, the conversion of the needle to the north; and therefore have studied to match and parallel these in the more obvious and neglected pieces of nature, which without further travel I can do in the cosmography of myself. We carry with us the wonders we seek without us: there is all Africa and her prodigies in us. We are that bold and
20 adventurous piece of nature which he that studies wisely learns in a compendium what others labour at in a divided piece and endless volume.

Section 16. Thus are there two books from whence I collect my divinity: besides that written one of God, another of his servant nature, that universal and public manuscript that lies expansed unto the eyes of all. Those that never saw him in the one have discovered him in the other. This was the scripture and theology of the heathens: the natural motion of the sun made them more admire him than its supernatural station
30 did the Children of Israel: the ordinary effects of nature wrought more admiration in them than in the other all his miracles. Surely the heathens knew better how to join and read these mystical letters than we Christians, who cast a more

careless eye on these common hieroglyphics, and disdain to suck divinity from the flowers of nature. Nor do I so forget God as to adore the name of nature; which I define not, with the Schools, the principle of motion and rest, but that straight and regular line, that settled and constant course the wisdom of God hath ordained the actions of his creatures, according to their several kinds. To make a revolution every day is the nature of the sun, because it is that necessary course which God hath ordained it, from which it cannot swerve but by a faculty from that voice which first did give it motion. Now this course of nature God seldom alters or perverts, but like an excellent artist hath so contrived his work that with the self-same instrument, without a new creation, he may effect his obscurest designs. Thus he sweetened the water with a wood; preserved the creatures in the Ark which the blast of his mouth might have as easily created: for God is like a skilful geometrician, who when more easily and with one stroke of his compass he might describe or divide a right line, had yet rather do this, though in a circle or longer way, according to the constituted and forelaid principles of his art. Yet this rule of his he doth sometimes pervert, to acquaint the world with his prerogative, lest the arrogancy of our reason should question his power and conclude he could not. And thus I call the effects of nature the works of God, whose hand and instrument she only is; and therefore to ascribe his actions unto her is to devolve the honour of the principal agent upon the instrument: which if with reason we may do, then let our hammers rise up and boast they have built our houses, and our pens receive the honour of our writings. I hold there is a general beauty in the works of God, and therefore no deformity in any kind or species of creature whatsoever. I cannot tell by what logic we call a toad, a bear or an elephant ugly; they being created in those outward shapes and figures which best express the actions of their inward forms, and having passed with

approbation that general visitation of God, who saw that all that he had made was good—that is, conformable to his will, which abhors deformity and is the rule of order and beauty. There is therefore no deformity but in monstrosity; wherein notwithstanding there is a kind of beauty, nature so ingeniously contriving those irregular parts as they become sometimes more remarkable than the principal fabric. To speak yet more narrowly, there was never anything ugly or unshapen but the Chaos; wherein notwithstanding—to speak strictly—
10 there was no deformity because no form, nor was it yet impregnate by the voice of God. Now, nature is not at variance with art, nor art with nature, they both being the servants of his providence: art is the perfection of nature. Were the world now as it was the sixth day, there were yet a Chaos: nature hath made one world, and art another. In brief, all things are artificial, for nature is the art of God.

Section 17. This is the ordinary and open way of his providence, which art and industry have in a good part discovered; whose effects we may foretell without an oracle. To foreshow
20 these is not prophecy but prognostication. There is another way, full of meanders and labyrinths, whereof the devil and spirits have no exact ephemerides, and that is a more particular and obscure method of his providence, directing the operations of individuals and single essences. This we call fortune, that serpentine and crooked line whereby he draws those actions that his wisdom intends in a more unknown and secret way. This cryptic and involved method of his providence have I ever admired; nor can I relate the history of my life, the occurrences of my days, the escapes of dangers and
30 hits of chance with a *bezo las manos* to fortune, or a bare gramercy to my good stars. Abraham might have thought the ram in the thicket came thither by accident; human reason would have said that mere chance conveyed Moses in the ark

to the sight of Pharaoh's daughter. What a labyrinth is there in the story of Joseph, able to convert a Stoic! Surely there are in every man's life certain rubs, doublings and wrenches which pass awhile under the effects of chance, but at the last, well examined, prove the mere hand of God. 'Twas not dumb chance that, to discover the Fougade or Powder-plot, contrived a miscarriage in the letter. I like the victory of '88 the better for that one occurrence which our enemies imputed to our dishonour and the partiality of fortune, to wit, the tempests and contrarieties of winds. King Philip did not detract from the nation when he said he sent his Armada to fight with men, and not to combat with the winds. Where there is a manifest disproportion between the powers and forces of two several agents, upon a maxim of reason we may promise the victory to the superior; but when unexpected accidents slip in, and unthought-of occurrences intervene, these must proceed from a power that owes no obedience to those axioms; where, as in the writing upon the wall, we behold the hand but see not the spring that moves it. The success of that petty province of Holland—of which the Grand Seignieur proudly said that if they should trouble him as they did the Spaniard, he would send his men with shovels and pickaxes, and throw it into the sea—I cannot altogether ascribe to the ingenuity and industry of the people, but to the mercy of God that hath disposed them to such a thriving genius, and to the will of providence that dispenseth her favour to each country in their preordinate season. All cannot be happy at once, for because the glory of one state depends upon the ruin of another, there is a revolution and vicissitude of their greatness, which must obey the swing of that wheel not moved by intelligences but by the hand of God, whereby all estates arise to their zenith and vertical points, according to their predestinated period. For the lives not only of men, but of commonweals and the whole world, run not upon an helix that still enlargeth but on a circle, where, arriving to their meridian, they decline in obscurity and fall under the horizon again.

Section 18. These must not therefore be named the effects of fortune but in a relative way, and as we term the works of nature. It was the ignorance of man's reason that begat this very name, and by a careless term miscalled the providence of God: for there is no liberty for causes to operate in a loose and straggling way, nor any effect whatsoever but hath its warrant
10 from some universal and more superior cause. 'Tis not a ridiculous devotion to say a prayer before a game at tables; for even in sortileges and matters of the greatest uncertainty there is a settled and preordered course of effects. 'Tis we that are blind, not fortune: because our eye is too dim to discover the mystery of her effects we foolishly paint her blind, and hoodwink the providence of the Almighty. I cannot justify that contemptible proverb that fools are only fortunate, nor that insolent paradox that a wise man is out of the reach of fortune; much less those opprobrious epithets of poets: whore,
20 bawd, and strumpet. 'Tis, I confess, the common fate of men of singular gifts of mind to be destitute of those of fortune; which doth not any way deject the spirits of wiser judgements, who rightly understand the justice of this proceeding; and being enriched with higher donatives, cast a more careless eye on these vulgar parts of felicity. 'Tis a most unjust ambition to desire to engross the mercies of the Almighty, nor to be content with the goods of mind without a possession of those of body and fortune; and 'tis an error worse than heresy to adore these complemental and circumstantial pieces of felicity, and
30 undervalue those perfections and essential points of happiness wherein we resemble our Maker. To wiser desires 'tis satisfaction enough to deserve, though not to enjoy, the favours of fortune. Let providence provide for fools: 'tis not partiality

but equity in God, who deals with us but as our natural parents. Those that are able in body and mind he leaves to their deserts; to those of weaker merits he imparts a larger portion, and pieces out the defect of one by the excess of the other. Thus have we no just quarrel with nature for leaving us naked, or to envy the horns, hooves, skins and furs of other creatures, being provided with reason that can supply them all. We need not labour with so many arguments to confute judicial astrology; for if there be a truth therein it doth not injure divinity. If to be born under Mercury disposeth us to be witty, under Jupiter to be wealthy, I do not owe a knee unto these, but unto that merciful hand that hath disposed and ordered my indifferent and uncertain nativity unto such benevolous aspects. Those that held that all things were governed by fortune had not erred, had they not persisted there. The Romans, that erected a temple to fortune, acknowledged therein, though in a blinder way, somewhat of divinity; for in a wise supputation all things begin and end in the Almighty. There is a nearer way to heaven than Homer's chain. An easy logic may conjoin heaven and earth in one argument, and with less than a sorites resolve all things into God. For though we christen effects by their most sensible and nearest causes, yet is God the true and infallible cause of all; whose concourse, though it be general, yet doth it subdivide itself into the particular actions of everything, and is that spirit by which each singular essence not only subsists but performs its operations.

Section 19. The bad construction and perverse comment on these pair of second causes, or visible hands of God, have perverted the devotion of many into atheism; who, forgetting the honest advisoes of faith, have listened unto the conspiracy of passion and reason. I have therefore always endeavoured to compose those feuds and angry dissensions between affection, faith and reason; for there is in our soul

a kind of triumvirate or triple government of three com-
petitors, which distract the peace of this our commonwealth
no less than did that other the state of Rome.

As reason is a rebel unto faith, so passion unto reason: as the
propositions of faith seem absurd unto reason, so the theorems
of reason unto passion, and both unto faith. Yet a moderate
and peaceable discretion may so state and order the matter
that they may be all kings and yet make but one monarchy,
everyone exercising his sovereignty and prerogative in a due
10 time and place, according to the restraint and limit of circumstance. There is, as in philosophy, so in divinity, sturdy doubts
and boisterous objections, wherewith the unhappiness of our
knowledge too nearly acquainteth us. More of these no man
hath known than myself, which I confess I conquered not in a
martial posture but on my knees. For our endeavours are not
only to combat with doubts, but always to dispute with the
devil. The villainy of that spirit takes a hint of infidelity from
our studies, and by demonstrating a naturality in one way
makes us mistrust a miracle in another. Thus, having perused
20 the Archidoxis and read the secret sympathies of things, he
would dissuade my belief from the miracle of the brazen
serpent, make me conceit that image worked by sympathy,
and was but an Egyptian trick to cure their diseases without
a miracle. Again, having seen some experiments of bitumen,
and having read far more of naphtha, he whispered to my
curiosity the fire of the altar might be natural, and bid me
mistrust a miracle in Elias when he entrenched the altar round
with water; for that inflammable substance yields not easily
unto water, but flames in the arms of its antagonist. And thus
30 would he inveigle my belief to think the combustion of Sodom
might be natural, and that there was an asphaltic and bituminous nature in that lake before the fire of Gomorrah.
I know that manna is now plentifully gathered in Calabria,
and Josephus tells me in his days 'twas as plentiful in Arabia.

The devil therefore made the query, Where was then the miracle in the days of Moses? The Israelites saw but that in his time the natives of those countries behold in ours. Thus the devil played at chess with me, and yielding a pawn, thought to gain a queen of me, taking advantage of my honest endeavours; and whilst I laboured to raise the structure of my reason, he strived to undermine the edifice of my faith.

Section 20. Neither had these, or any other, ever such advantage of me as to incline me to any point of infidelity or desperate positions of atheism, for I have been these many years of 10 opinion there was never any. Those that held religion was the difference of man from beasts have spoken probably, and proceed upon a principle as inductive as the other. That doctrine of Epicurus that denied the providence of God was no atheism but a magnificent and high-strained conceit of his majesty, which he deemed too sublime to mind the trivial actions of those inferior creatures. That fatal necessity of the Stoics is nothing but the immutable law of his will. Those that heretofore denied the divinity of the Holy Ghost have been condemned but as heretics; and those that now deny our Saviour— 20 though more than heretics—are not so much as atheists; for though they deny two persons in the Trinity, they hold, as we do, there is but one God.

That villain and secretary of hell that composed that miscreant piece of the three impostors, though divided from all religions, and was neither Jew, Turk nor Christian, was not yet a positive atheist. I confess every country hath its Machiavel, every age its Lucian, whereof common heads must not hear, nor more advanced judgements too rashly venture on. 'Tis the rhetoric of Satan, and may pervert a loose or prejudicate belief. 30

Section 21. I confess I have perused them all, and can discover nothing that may startle a discreet belief; yet are there heads

carried off with the wind and breath of such motives. I
remember a doctor in physic of Italy who could not perfectly
believe the immortality of the soul because Galen seemed to
make a doubt thereof. With another I was familiarly acquainted in France, a divine and a man of singular parts, that
on the same point was so plunged and gravelled with three
lines of Seneca that all our antidotes, drawn from both
Scripture and philosophy, could not expel the poison of his
error. There are a set of heads that can credit the relations of
10 mariners yet question the testimony of St Paul; and peremptorily maintain the traditions of Aelian or Pliny yet in histories
of Scripture raise queries and objections, believing no more
than they can parallel in human authors. I confess there are
in Scripture stories that do exceed the fables of poets, and to
a captious reader sound like Gargantua or Bevis. Search all
the legends of times past and the fabulous conceits of these
present, and 'twill be hard to find one that deserves to carry
the buckler unto Samson; yet is all this of an easy possibility
if we conceive a divine concourse or an influence but from the
20 little finger of the Almighty. It is impossible that either in the
discourse of man or in the infallible voice of God, to the weakness of our apprehension, there should not appear irregularities,
contradictions and antinomies. Myself could show a catalogue of doubts never yet imagined nor questioned by any, as
I know, which are not resolved at the first hearing; not queries
fantastic or objections of air: for I cannot hear of atoms in
divinity. I can read the story of the pigeon that was sent out
of the Ark and returned no more, yet not question how she
found out her mate that was left behind: that Lazarus was
30 raised from the dead, yet not demand where in the interim his
soul awaited; or raise a law-case whether his heir might law-
fully detain his inheritance, bequeathed unto him by his death,
and he, though restored to life, have no plea or title unto his
former possessions. Whether Eve was framed out of the left

side of Adam I dispute not, because I stand not yet assured which is the right side of a man, or whether there be any such distinction in nature. That she was edified out of the rib of Adam I believe, yet raise no question who shall arise with that rib at the Resurrection: whether Adam was an hermaphrodite, as the rabbins contend upon the letter of the text, because it is contrary to all reason that there should be an hermaphrodite before there was a woman, or a composition of two natures before there was a second composed. Likewise, whether the world was created in autumn, summer or spring, because it 10 was created in them all; for whatsoever sign the sun possesseth, those four seasons are actually existent. It is the nature of this luminary to distinguish the several seasons of the year, all which it makes at one time in the whole earth, and successively in any part thereof. There are a bundle of curiosities, not only in philosophy but in divinity, proposed and discussed by men of the most supposed abilities, which indeed are not worthy our vacant hours, much less our more serious studies: pieces only fit to be placed in Pantagruel's library, or bound up with Tartaretus' *De modo cacandi*. 20

Section 22. These are niceties that become not those that peruse so serious a mystery. There are others more generally questioned and called to the bar, yet methinks of an easy and possible truth. 'Tis ridiculous to put off or drown the general Flood of Noah in that particular inundation of Deucalion. That there was a deluge once seems not to me so great a miracle as that there is not one always. How all the several kinds of creatures, not only in their own bulks but with a competency of food and sustenance, might be preserved in one Ark and within the extent of three hundred cubits, to a reason 30 that rightly examines it will appear very feasible. There is another secret not contained in the Scripture which is more hard to comprehend, and put the honest Father to the refuge

27

of a miracle; and that is, not only how the distinct pieces of the world and divided islands should be first planted by men, but inhabited by tigers, panthers and bears. How America abounded with beasts of prey and noxious animals, yet contained not in it that necessary creature a horse, is very strange. By what passage those not only birds but dangerous and unwelcome beasts came over; how there be creatures there which are not found in this triple continent: all which must needs be strange unto us that hold but one Ark, and that the
10 creatures began their progress from the mountain of Ararat. They who, to salve this, would make the deluge particular, proceed upon a principle that I can no way grant; not only upon the negative of holy Scriptures but of mine own reason, whereby I can make it probable that the world was as well peopled in the time of Noah as in ours, and fifteen hundred years to people the world as full a time for them as four thousand years since have been to us. There are other assertions and common tenets drawn from Scripture and generally believed as Scripture, whereunto notwithstanding I would
20 never betray the liberty of my reason. 'Tis a postulate to me that Methuselah was the longest lived of all the children of Adam; and no man will be able to prove it, when from the process of the text I can manifest it may be otherwise. That Judas perished by hanging himself there is no certainty in Scripture. Though in one place it seems to affirm it, and by a doubtful word hath given occasion so to translate it, yet in another place, in a more punctual description, it makes it improbable and seems to overthrow it. That our fathers, after the Flood, erected the Tower of Babel to preserve themselves
30 against a second deluge, is generally opinioned and believed; yet is there another intention of theirs expressed in Scripture. Besides, it is improbable from the circumstance of the place, that is, a plain in the land of Shinar. These are no points of faith, and therefore may admit a free dispute. There are yet others,

28

and those familiarly concluded from the text, wherein—under favour—I see no consequence. To instance in one or two, as to prove the Trinity from that speech of God in the plural number, *faciamus hominem*, let us make man—which is but the common style of princes and men of eminency—he that shall read one of his Majesty's proclamations may with the same logic conclude there be two kings in England. To infer the obedient respect of wives to their husbands from the example of Sarah, who usually called her husband Lord; which if you examine, you shall find to be no more than Signior or Monsieur, which are the ordinary languages all civil nations use in their familiar compellations, not to their superiors or equals but to their inferiors also, and persons of lower condition. The Church of Rome confidently proves the opinion of tutelary angels from that answer when Peter knocked at the door, '*Tis not he, but his angel*; that is, might some say, his messenger or somebody from him; for so the original signifies, and is as likely to be the doubtful familist's meaning. This exposition I once suggested to a young divine that answered upon this point; to which, I remember, the Franciscan opponent replied no more but that it was a new and no authentic interpretation.

Section 23. These are but the conclusions and fallible discourses of man upon the word of God, for such do I believe the holy Scriptures: yet, were it of man, I could not choose but say it were the singularest and superlative piece that hath been extant since the Creation. Were I a pagan, I should not refrain the lecture of it; and cannot but commend the judgement of Ptolemy, that thought not his library complete without it. The Alcoran of the Turks—I speak without prejudice—is an ill-composed piece, containing in it vain and ridiculous errors in philosophy, superfluous repetitions, impossibilities, fictions and vanities beyond laughter, maintained

by evident and open sophisms; the policy of ignorance, deposition of universities and banishment of learning, that hath
gotten foot by arms and violence. This without a blow hath
disseminated itself through the whole earth. It is not un-
remarkable what Philo first observed, that the law of Moses
continued two thousand years without the least alteration,
whereas we see the laws of other commonweals to alter with
occasions; and even those that pretended their original from
some divinity to have vanished without trace or memory.
10 I believe besides Zoroaster there were divers that wrote before
Moses, who notwithstanding have suffered the common fate
of time. Men's works have an age like themselves; and though
they outlive their authors, have yet a stint and period of their
duration. This only is a work too hard for the teeth of time,
and cannot perish but in those general flames when all things
shall confess their ashes.

Section 24. I have heard some with deep sighs lament the lost
lines of Cicero; others with as many groans deplore the combustion of the library of Alexandria. For my own part I think
20 there be too many in the world, and could with patience
behold the urn and ashes of the Vatican could I, with a few
others, recover the perished leaves of Solomon, the sayings of
the Seers and the chronicles of the kings of Judah. I would not
omit a copy of Enoch's Pillars, had they many nearer authors
than Josephus, or did not relish somewhat of the fable. Some
men have written more than others have spoken. Pineda quotes
more authors in one work than are necessary in a whole world.
Of those three great inventions of Germany there are two
which are not without their incommodities, and 'tis disputable
30 whether they exceed not their use and commodities. 'Tis not
a melancholy utinam of mine own, but the desire of better
heads that there were a general synod; not to unite the incompatible difference of religion, but for the benefit of learning, to

reduce it as it lay at first in a few and solid authors, and to condemn to the fire those swarms and millions of rhapsodies, begotten only to distract and abuse the weaker judgements of scholars, and to maintain the trade and mystery of typographers.

Section 25. I cannot but wonder with what exceptions the Samaritans could confine their belief to the Pentateuch or five books of Moses. I am amazed at the rabbinical interpretations of the Jews upon the Old Testament as much as their defection from the New; and truly it is beyond wonder how that contemptible and degenerate issue of Jacob, once so devoted to ethnic superstition and so easily seduced to the idolatry of their neighbours, should now in such an obstinate and peremptory belief adhere unto their own doctrine, expect impossibilities, and in the face and eye of the Church persist without the least hope of conversion. This is a vice in them, but were a virtue in us; for obstinacy in a bad cause is but constancy in a good. And herein I must accuse my own religion; for there is not any of such a fugitive faith, such an unstable belief, as a Christian; none that do so oft transform themselves, not unto several shapes of Christianity and of the same species, but unto more unnatural and contrary forms of Jew and Mahometan, that from the name of Saviour can descend to the bare term of Prophet; and from an old belief that he is come, fall to a new expectation of his coming. It is the promise of Christ to make us all one flock; but how and when this union shall be is as obscure to me as the last day. Of those four members of religion we hold a slender proportion. There are, I confess, some new additions, yet small to those which accrue to our adversaries, and those only drawn from the revolt of pagans; men but of negative impieties, and such as deny Christ but because they never heard of him. But the religion of the Jew is expressly against the Christian, and the Mahometan against

both; for the Turk, in the bulk he now stands, is beyond all hope of conversion. If he fall asunder there may be conceived hopes, but not without strong improbabilities. The Jew is obstinate in all fortunes. The persecution of fifteen hundred years hath but confirmed them in their error: they have already endured whatsoever may be inflicted, and have suffered, in a bad cause, even to the commendation of their enemies. Persecution is a bad and indirect way to plant religion. It hath been the unhappy method of angry devo-
10 tions not only to confirm honest religion but wicked heresies and extravagant opinions. It was the first stone and basis of our faith: none can more justly boast of persecutions or glory in the number and nature of our martyrs; for to speak properly, these are the true and almost only examples of forti-tude. Those that are fetched from the field or drawn from the actions of the camp are not ofttimes so truly precedents of valour as audacity, and at the best attain but to some bastard piece of fortitude. If we shall strictly examine the circum-stances and requisites which Aristotle requires to true and
20 perfect valour, we shall find the name only in his master, Alexander, and as little in that Roman worthy, Julius Caesar; and if any in that easy and active part have done so nobly as to deserve that name, yet in the passive and more terrible piece these have surpassed, and in a more heroical way may claim the honour of that title. 'Tis not in the power of every honest faith to proceed thus far, or pass to heaven through the flames. Everyone hath it not in that full measure, nor in so audacious and resolute a temper as to endure those terrible tests and trials; who notwithstanding in a peaceable way do truly adore
30 their Saviour, and have, no doubt, a faith acceptable in the eyes of God.

Section 26. Now as all that die in war are not termed soldiers, so neither can I properly term all those that suffer in matters of religion martyrs. The Council of Constance condemns John Huss for an heretic; the stories of his own party style him a martyr. He must needs offend the divinity of both that says he was neither the one nor the other. There are, questionless, many canonised on earth that shall never be called saints in heaven, and have their names in histories and martyrologies who in the eyes of God are not so perfect martyrs as was that wise heathen Socrates, that suffered on a fundamental point of religion, the unity of God. I have often pitied the miserable bishop that suffered in the cause of Antipodes, yet cannot 10 choose but accuse him of as much madness for exposing his living on such a trifle as those of ignorance and folly that condemned him. I think my conscience will not give me the lie if I say there are not many extant that in a noble way fear the face of death less than myself; yet from the moral duty I owe to the commandments of God, and the natural respect that I tender unto the conservation of my essence and being, I would not perish upon a ceremony, politic point or indifferency; nor is my belief of that untractable temper as not to bow at their obstacles, or connive at matters wherein there are not 20 manifest impieties. The leaven, therefore, and ferment of all— not only civil but religious actions—is wisdom; without which, to commit ourselves to the flame is homicide, and, I fear, but to pass through one fire into another.

Section 27. That miracles are ceased I can neither prove nor absolutely deny, much less define the time and period of their cessation. That they survived Christ is manifest upon record of Scripture; that they outlived the Apostles also, and were revived at the conversion of nations many years after, we cannot deny, if we shall not question those writers whose 30 testimonies we do not controvert in points that make for our own opinions. Therefore that may have some truth in it that is reported of the Jesuits and their miracles in the Indies. I could

33

wish it were true, or had any other testimony than their own pens. They may easily believe those miracles abroad who daily conceive a greater at home: the transmutation of those visible elements into the body and blood of our Saviour: for the conversion of water into wine which he wrought in Cana, or what the devil would have him have done in the wilderness, of stones into bread, compared to this will scarce deserve the name of a miracle. Though indeed, to speak strictly, there is not one miracle greater than another, they being the extra-
10 ordinary effects of the hand of God, to which all things are of an equal facility, and to create the world as easy as one single creature. For this is also a miracle, not only to produce effects against or above nature, but before nature; and to create nature as great a miracle as to contradict or transcend her. We do too narrowly define the power of God, restraining it to our own capacities. I hold that God can do all things. How he should work contradictions I do not understand, yet dare not therefore deny. I cannot see why the Angel of God should question Esdras to recall the time past if it were beyond his
20 own power; or that God should pose mortality in that which he could not perform himself. I will not say God cannot, but he will not perform many things which we plainly affirm he cannot: this, I am sure, is the mannerliest proposition, wherein notwithstanding I hold no paradox. For strictly his power is but the same with his will, and they both, with all the rest, do make but one God.

Section 28. Therefore that miracles have been I do believe; that they may yet be wrought by the living I do not deny; but have no confidence in those which are fathered on the
30 dead: and this hath ever made me suspect the efficacy of relics, to examine the bones, question the habits and appurtenances of saints, and even of Christ himself. I cannot conceive why the cross that Helena found and whereon Christ himself died

34

should have power to restore others unto life: I excuse not Constantine from a fall off his horse or a mischief from his enemies upon the wearing those nails on his bridle which our Saviour bore upon the cross in his hands: I compute among your *piae fraudes*, nor many degrees before consecrated swords and roses, that which Baldwin, King of Jerusalem, returned the Genoese for their cost and pains in his war—to wit, the ashes of John the Baptist. Those that hold the sanctity of their souls doth leave behind a tincture and sacred faculty on their bodies speak naturally of miracles and do not salve the doubt. 10 Now one reason I tender so little devotion unto relics is, I think, the slender and doubtful respect I have always held unto antiquities: for that indeed which I admire is far before antiquity; that is, eternity, and that is God himself; who though he be styled the Ancient of Days, cannot receive the adjunct of antiquity, who was before the world and shall be after it, yet is not older than it. For in his years there is no climacter; his duration is eternity, and far more venerable than antiquity.

Section 29. But above all the rest I wonder how the curiosity of wiser heads could pass that great and indisputable miracle 20 of the cessation of oracles; and in what swoon their reasons lay to content themselves and sit down with such far-fetched and ridiculous reasons as Plutarch allegeth for it. The Jews, that can believe that supernatural solstice of the sun in the days of Joshua, have yet the impudence to deny the eclipse which every pagan confessed at his death: but for this, it is evident beyond contradiction: the devil himself confessed it. Certainly it is not a warrantable curiosity to examine the verity of Scripture by the concordance of human history, or seek to confirm the chronicle of Daniel or Esther by the authority of 30 Megasthenes or Herodotus. I confess I have had an unhappy curiosity this way, till I laughed myself out of it with a piece of Justin, where he delivers that the Children of Israel for being scabbed were banished Egypt. And truly, since I have understood the occurrences of the world, and know in what counterfeit shapes and deceitful vizards times present represent on the stage things past, I do believe them little more than things to come. Some have been of my opinion, and endeavoured to write the history of their own lives; wherein Moses hath outgone them all, and left not only the story of his life, but—as some will have it—of his death also.

Section 30. It is a riddle to me how this very story of oracles
10 hath not wormed out of the world that doubtful conceit of spirits and witches; how so many learned heads should so far forget their metaphysics, and destroy the ladder and scale of creatures as to question the existence of spirits. For mine own part I have ever believed, and do now know, that there are witches. They that doubt of these do not only deny them, but spirits; and are obliquely and upon consequence a sort not of infidels but atheists. Those that to confute their incredulity desire to see apparitions shall questionless never behold any, nor have the power to be so much as witches; the devil hath
20 them already in a heresy as capital as witchcraft, and to appear to them were but to convert them. Of all the delusions wherewith he deceives mortality, there is not any that puzzleth me more than the legerdemain of changelings. I do not credit those transformations of reasonable creatures into beasts, or that the devil hath a power to transpeciate a man into a horse, who tempted Christ—as a trial of his divinity—to convert but stones into bread. I could believe that spirits use with men the act of carnality, and that in both sexes: I conceive they may assume, steal or contrive a body wherein there may be action
30 enough to content decrepit lust, or passion to satisfy more active veneries, yet in both without a possibility of generation: and therefore that opinion that Antichrist should be born of the Tribe of Dan by the conjunction with the devil is ridicu-

lous, and a conceit fitter for a rabbin than a Christian. I hold that the devil doth really possess some men, the spirit of melancholy others, the spirit of delusion others; that as the devil is concealed and denied by some, so God and good angels are pretended by others, whereof the late detection of the Maid of Germany hath left a pregnant example.

Section 31. Again, I believe that all that use sorceries, incantations and spells are not witches or, as we term them, magicians. I conceive there is a traditional magic, not learned immediately from the devil but at second hand from his scholars; who having once the secret betrayed, are able and do empirically practise without his advice; they both proceeding upon the principles of nature, where actives, aptly conjoined to disposed passives, will under any master produce their effects. Thus I think a great part of philosophy was at first witchcraft, which being afterward derived to one another, proved but philosophy, and was indeed no more than the honest effects of nature. What invented by us is philosophy, learned from him is magic. We do surely owe the honour of many secrets to the discovery of good and bad angels. I could never pass that sentence of Paracelsus without an asterisk or annotation; *Ascendens constellatum multa revelat quaerentibus magnalia naturae,* (i.e. *opera Dei*). I do think that many mysteries ascribed to our own inventions have been the courteous revelations of spirits; for those noble essences in heaven bear a friendly regard unto their fellow natures on earth; and therefore believe that those many prodigies and ominous prognostics which forerun the ruins of states, princes and private persons are the charitable premonitions of good angels, which more careless enquiries term but the effects of chance and nature.

Section 32. Now besides these particular and divided spirits there may be, for aught I know, a universal and common

37

spirit to the whole world. It was the opinion of Plato, and it is yet of the Hermetical philosophers. If there be a common nature that unites and ties the scattered and divided individuals into one species, why may there not be one that unites them all? How soever, I am sure there is a common spirit that plays within us yet makes no part of us; and that is the spirit of God, the fire and scintillation of that noble and mighty essence which is the life and radical heat of spirits and those essences that know not the virtue of the sun; a fire quite con-
10 trary to the fire of hell. This is that gentle heat that brooded on the waters and in six days hatched the world; this is that irradiation that dispels the mists of hell, the clouds of horror, fear, sorrow, despair, and preserves the region of the mind in serenity. Whosoever feels not the warm gale and gentle venti-lation of this spirit, though I feel his pulse I dare not say he lives; for truly, without this, to me there is no heat under the tropic, nor any light, though I dwelt in the body of the sun.

> As, when the labouring sun hath wrought his track
> Up to the top of lofty Cancer's back,
> 20 The icy ocean cracks, the frozen pole
> Thaws with the heat of that celestial coal,
> So, when thy absent beams begin t'impart
> Again a solstice on my frozen heart,
> My winter's o'er, my drooping spirits sing,
> And every part revives into a spring.
> But if thy quick'ning beams awhile decline,
> And with their light bless not this orb of mine,
> A chilly frost surpriseth every member,
> And in the midst of June I find December.
> 30 Keep still in my horizon, for to me
> 'Tis not the sun that makes the day, but thee.
> O how this earthy temper doth debase
> The noble soul in this her humble place;
> Whose wingy nature ever doth aspire
> To reach that place whence first it took its fire.

These flames I feel—which in my heart do dwell—
Are not thy beams, but take their fire from hell:
O quench them all, and let thy light divine
Be e'er the sun to this poor orb of mine;
And to thy sacred spirit convert those fires
Whose earthy fumes choke my devout aspires.

Section 33. Therefore for spirits, I am so far from denying their existence that I could easily believe that not only whole countries but particular persons have their tutelary and guardian angels. It is not a new opinion of the Church of Rome, but an old one of Pythagoras and Plato. There is no heresy in it, and if not manifestly defined in Scripture yet is it an opinion of a good and wholesome use in the course and actions of a man's life, and would serve as an hypothesis to salve many doubts whereof common philosophy affordeth no solution. Now if you demand my opinion and metaphysics of their natures, I confess them very shallow; most in a negative way, like that of God; or in a comparative, between ourselves and fellow-creatures; for there is in this universe a stair or manifest scale of creatures, rising not disorderly or in confusion but with a comely method and proportion. Between creatures of mere existence and things of life there is a large disproportion of nature; between plants and animals or creatures of sense, a wider difference; between them and man, a far greater: and if the proportion hold on, between man and angels there should be yet a greater. We do not comprehend their natures, who retain the definition of Porphyry and distinguish them from ourselves by immortality; for before his fall, 'tis thought, man also was immortal; yet must we needs affirm that he had a different essence from the angels. Having therefore no certain knowledge of their natures, 'tis no bad method of the Schools, whatsoever perfection we find obscurely in ourselves, in a more complete

and absolute way to ascribe unto them. I believe they have
an extemporary knowledge, and upon the first motion of their
reason do what we cannot without study or deliberation;
that they know things by their forms, and define by specifical
differences what we describe by accidents and properties—
and therefore probabilities to us may be demonstrations unto
them; that they have knowledge not only of the specifical
but the numerical forms of individuals, and understand by
what reserved difference each single hypostasis—besides the
10 relation to its species—becomes its numerical self; that as the
soul hath a power to move the body it informs, so theirs
a faculty to move any, though inform none; ours upon
restraint of time, place and distance. But that invisible hand
that conveyed Habakkuk to the lion's den, or Philip to
Azotus, infringeth this rule, and hath a secret conveyance
wherewith mortality is not acquainted. If they have that
intuitive knowledge whereby as in reflection they behold the
thoughts of one another, I cannot peremptorily deny but they
know a great part of ours. They that to refute the invocation
20 of saints have denied that they have any knowledge of our
affairs below have proceeded too far, and must pardon my
opinion till I can thoroughly answer that piece of Scripture,
At the conversion of a sinner, the angels of heaven rejoice. I cannot,
with those, in that great Father securely interpret the work of
the first day, *Fiat lux,* to the creation of angels; though I confess there is not any creature that hath so near a glimpse of their
nature as light in the sun and elements. We style it a bare
accident, but where it subsists alone 'tis a spiritual substance,
and may be an angel: in brief, conceive light invisible, and
30 that is a spirit.

Section 34. These are certainly the magisterial and master-
pieces of the Creator; the flower, or—as we may say—the best
part of nothing, actually existing; what we are but in hopes

and probability. We are only that amphibious piece between a corporal and spiritual essence; that middle frame that links those two together, and makes good the method of God and nature, that jumps not from extremes but unites the incompatible distances by some middle and participating natures. That we are the breath and similitude of God, it is indisputable and upon record of holy Scripture; but to call ourselves a microcosm or little world, I thought it only a pleasant trope of rhetoric till my nearer judgement and second thoughts told me there was a real truth therein. For first we are a rude mass, and in the rank of creatures which only are, and have a dull kind of being not yet privileged with life, or preferred to sense or reason. Next we live the life of plants, the life of animals, the life of men, and at last the life of spirits; running on, in one mysterious nature, those five kinds of existences which comprehend the creatures not of the world only, but of the universe. Thus is man that great and true amphibium, whose nature is disposed to live not only like other creatures in divers elements, but in divided and distinguished worlds. For though there be but one world to sense, there are two to reason; the one visible, the other invisible, whereof Moses seems to have left no description; and of the other so obscurely that some parts thereof are yet in controversy: and truly, for the first chapters of Genesis I must confess a great deal of obscurity. Though divines have, to the power of human reason, endeavoured to make all go in a literal meaning, yet those allegorical interpretations are also probable, and perhaps the mystical method of Moses bred up in the hieroglyphical schools of the Egyptians.

Section 35. Now, for that immaterial world, methinks we need not wander so far as beyond the first moveable; for even in this material fabric the spirits walk as freely exempt from the affections of time, place and motion as beyond the extremest

42

circumference. Do but extract from the corpulency of bodies, or resolve things beyond their first matter, and you discover the habitation of angels; which if I call the ubiquitary and omnipresent essence of God I hope I shall not offend divinity; for before the creation of the world God was really all things. For the angels he created no new world or determinate mansion, and therefore they are everywhere where is his essence, and do live, at a distance even, in himself. That God made all things for man is in some sense true, yet not so far as to subordinate the creation of those purer creatures unto ours; though as ministering spirits they do and are willing to fulfil the will of God in these lower and sublunary affairs of man. God made all things for himself, and it is impossible he should make them for any other end than his own glory. It is all he can receive, and all that is without himself; for honour being an external adjunct, and in the honourer rather than in the person honoured, it was necessary to make a creature from whom he might receive this homage; and that is, in the other world angels, in this, man: which when we neglect we forget the very end of our creation, and may justly provoke God not only to repent that he hath made the world, but that he hath sworn he would not destroy it. That there is but one world is a conclusion of faith: Aristotle with all his philosophy hath not been able to prove it, and as weakly that the world was eternal. That dispute much troubled the pens of ancient philosophers, who saw no further than the first matter; but Moses hath decided that question, and all is salved with the new term of a creation, that is, a production of something out of nothing. And what is that? whatsoever is opposite to something; or more exactly, that which is truly contrary unto God: for he only is, all others have an existence with dependency, and are something but by a distinction. And herein is divinity conformant unto philosophy, and generation not only founded on contrarieties, but also creation. God, being

43

all things, is contrary unto nothing, out of which were made all things; and so nothing becomes something, and omneity informed nullity into an essence.

Section 36. The whole creation is a mystery, and particularly that of man. At the blast of his mouth were the rest of the creatures made, and at his bare word they started out of nothing; but in the frame of man—as the text describes it—he played the sensible operator, and seemed not so much to create as make him. When he had separated the materials of
10 other creatures there consequently resulted a form and soul, but having raised the walls of man he was driven to a second and harder creation, of a substance like himself, an incorruptible and immortal soul. For these two affections we have the philosophy and opinion of the heathens, the flat affirmative of Plato and not a negative from Aristotle. There is another scruple cast in by divinity concerning its production, much disputed in the German auditories, and with that indifferency and equality of arguments as leave the controversy undetermined. I am not of Paracelsus' mind, that boldly delivers a receipt to
20 make a man without conjunction, yet cannot but wonder at the multitude of heads that do deny traduction, having no other argument to confirm their belief than that rhetorical sentence and antimetathesis of Augustine, *Creando infunditur, infundendo creatur.* Either opinion will consist well enough with religion, yet I should rather incline to this did not one objection haunt me, not wrung from speculations and subtleties, but from common sense and observation; not picked from the leaves of any author, but bred amongst the weeds and tares of mine own brain. And this is a conclusion from the
30 equivocal and monstrous productions in the copulation of man with beast; for if the soul be not transmitted and transfused in the seed of the parents, why are not these productions merely beasts, but have also a tincture and impression of

reason in as high a measure as it can evidence itself in those
improper organs? Nor truly can I peremptorily deny that the
soul, in this her sublunary estate, is wholly and in all acceptions
inorganical, but that for the performance of her ordinary
actions there is required not only a symmetry and proper
disposition of organs, but a crasis and temper correspondent to
its operation. Yet is not this mass of flesh and visible structure
the instrument and proper corps of the soul, but rather of
sense, and that the hand of reason. In our study of anatomy
there is a mass of mysterious philosophy, and such as reduced 10
the very heathens to divinity; yet amongst all those rare discoveries and curious pieces I find in the fabric of man I do not
so much content myself as in that I find not; that is, no organ
or proper instrument for the rational soul. For in the brain,
which we term the seat of reason, there is not anything of
moment more than I can discover in the cranie of a beast: and
this is a sensible and no inconsiderable argument of the inorganity of the soul, at least in that sense we usually so receive
it. Thus are we men, and we know not how; there is something in us that can be without us, and will be after us; though 20
it is strange that it hath no history what it was before us, nor
can tell how it entered in us.

Section 37. Now, for the walls of flesh wherein the soul doth
seem to be immured before the Resurrection, it is nothing
but an elemental composition and a fabric that must fall to
ashes. *All flesh is grass* is not only metaphorically but literally
true, for all those creatures which we behold are but the herbs
of the field digested into flesh in them, or more remotely
carnified in ourselves. Nay, further, we are what we all
abhor, anthropophagi and cannibals, devourers not only of 30
men but of ourselves, and that not in an allegory but a positive
truth. For all this mass of flesh that we behold came in at our
mouths; this frame we look upon hath been upon our

trenchers. In brief, we have devoured ourselves, and yet do live and remain ourselves. I cannot believe the wisdom of Pythagoras did ever positively and in a literal sense affirm his metempsychosis or impossible transmigration of the souls of men into beasts. Of all metamorphoses and transformations I believe only one; that is, of Lot's wife, for that of Nebuchadnezzar proceeded not so far. In all others I conceive no further verity than is contained in their implicit sense and morality. I believe that the whole frame of a beast doth
10 perish, and is left in the same estate after death as before it was materialled into life; that the souls of men know neither contrary nor corruption; that they subsist beyond the body, and outlive death by the privilege of their proper natures, and without a miracle; that the souls of the faithful, as they leave earth, take possession of heaven; that those apparitions and ghosts of departed persons are not the wandering souls of men but the unquiet walks of devils, prompting and suggesting us unto mischief, blood and villainy; instilling and stealing into our hearts that the blessed spirits are not at rest in their graves,
20 but wander solicitous of the affairs of the world. That these phantasms appear often, and do frequent cemeteries, charnel houses and churches, it is because these are the dormitories of the dead, where the devil, like an insolent champion, beholds with pride the spoils and trophies of his victory in Adam.

Section 38. This is that dismal conquest we all deplore, that makes us so often cry, *O Adam, quid fecisti?* I thank God I have not those strait ligaments or narrow obligations unto the world as to dote on life, or be convulsed and tremble at the name of death. Not that I am insensible of the dread and
30 horror thereof, or by raking into the bowels of the deceased, or the continual sight of anatomies, skeletons or cadaverous relics, like vespilloes or grave-makers, I am become stupid or have forgot the apprehension of mortality; but that marshal-

ling all the horrors and contemplating the extremities thereof, I find not anything therein able to daunt the courage of a man, much less a well-resolved Christian. And therefore am not angry with the error of our first parents or unwilling to bear a part of this common fate, and like the best of them, to die: that is, to cease to breathe, to take a farewell of the elements, to be a kind of nothing for a moment, to be within one instant of a spirit. When I take a full view and circle of myself without this reasonable moderator and equal piece of justice, death, I do conceive myself the miserablest person extant. Were there not another life that I hope for, all the vanities of this world should not entreat a moment's breath from me. Could the devil work my belief to imagine I could never die, I would not outlive that very thought. I have so abject a conceit of this common way of existence, this retaining to the sun and elements, I cannot think this is to be a man or to live according to the dignity of humanity. In expectation of a better, I can with patience embrace this life, yet in my best meditations do often desire death. It is a symptom of melancholy to be afraid of death, yet sometimes to desire it. This latter I have often discovered in myself, and think no man ever desired life as I have sometimes death. I honour any man that contemns it, nor can I highly love any that is afraid of it. This makes me naturally love a soldier, and honour those tattered and contemptible regiments that will die at the command of a sergeant. For a pagan there may be motives to be in love with life; but for a Christian that is amazed at death, I see not how he can escape this dilemma: that he is too sensible of this life, or hopeless of the life to come.

Section 39. Some divines count Adam thirty years old at his creation, because they suppose him created in the perfect age and stature of man; and surely we are all out of the computation of our age, and every man is some months older than he

bethinks him. For we live, move, have a being and are subject to the actions of the elements and the malice of diseases in that other world, the truest microcosm, the womb of our mother; for besides that general and common existence we are conceived to hold in our chaos, and whilst we sleep within the bosom of our causes, we enjoy a being and life in three distinct worlds, wherein we receive most manifest graduations. In that obscure world and womb of our mother our time is short, computed by the moon; yet longer than
10 the days of many creatures that behold the sun; ourselves being not yet without life, sense and reason, though for the manifestation of its actions it awaits the opportunity of objects, and seems to live there but in its root and soul of vegetation. Entering afterwards upon the scene of the world, we arise up and become another creature, performing the reasonable actions of man and obscurely manifesting that part of divinity in us, but not in complement and perfection till we have once more cast our secundine—that is, this slough of flesh—and are delivered into the last world, that ineffable place of Paul, that
20 proper *ubi* of spirits. The smattering I have of the philosopher's stone—which is nothing else but the perfectest exaltation of gold—hath taught me a great deal of divinity, and instructed my belief how that immortal spirit and incorruptible substance of my soul may lie obscure and sleep awhile within this house of flesh. Those strange and mystical transmigrations that I have observed in silkworms turned my philosophy into divinity. There is in those works of nature which seem to puzzle reason something divine, and that hath more in it than the eye of a common spectator doth discover. I have therefore
30 forsaken those strict definitions of death by privation of life, extinction of natural heat, separation etc., of soul and body, and have framed one in an hermetical way unto mine own fancy: *Est mutatio ultima qua perficitur nobile illud extractum microcosmi*. For to me, that consider things in a natural and

experimental way, man seems to be but a digestion or a preparative way unto that last and glorious elixir which lies imprisoned in the chains of flesh.

Section 40. I am naturally bashful, nor hath conversation, age or travel been able to effront or enharden me, yet I have one part of modesty which I have seldom discovered in another: that is—to speak truly—I am not so much afraid of death as ashamed thereof. 'Tis the very disgrace and ignominy of our natures, that in a moment can so disfigure us that our nearest friends, wife and children stand afraid and start at us. The birds 10 and beasts of the field, that before in a natural fear obeyed us, forgetting all allegiance begin to prey upon us. This very conceit hath in a tempest disposed and left me willing to be swallowed in the abyss of waters, wherein I had perished unseen, unpitied, without wondering eyes, tears of pity, lectures of mortality; and none had said, *Quantum mutatus ab illo!* Not that I am ashamed of the anatomy of my parts, or can accuse nature for playing the bungler in any part of me, or my own vicious life for contracting any shameful disease upon me whereby I might not call myself as wholesome a 20 morsel for the worms as any.

Section 41. Some, upon the courage of a fruitful issue, wherein, as in the truest chronicle, they seem to outlive themselves, can with greater patience away with death. This conceit and counterfeit subsisting in our progenies seems to me a mere fallacy, unworthy the desires of a man that can but conceive a thought of the next world, who in a nobler ambition should desire to live in his substance in heaven rather than in his name and shadow on earth. And therefore at my death I mean to take a total adieu of the world, not caring for a monument, 30 history or epitaph; not so much as the bare memory of my name to be found anywhere but in the universal register of

God. I am not yet so cynical as to approve the testament of Diogenes, nor do I altogether allow that rodomontado of Lucan,

Coelo tegitur, qui non habet urnam.
He that unburied lies wants not his hearse,
For unto him a tomb's the universe

but commend in my calmer judgement their ingenuous intentions that desire to sleep by the urns of their fathers, and strive to go the nearest way unto corruption. I do not envy the temper of crows and daws, nor the numerous and weary days of our fathers before the Flood. If there be any truth in astrology, I may outlive a jubilee. As yet I have not seen one revolution of Saturn, nor hath my pulse beat thirty years, and yet, excepting one, have seen the ashes and left underground all the kings of Europe; have been contemporary to three emperors, four Grand Seignieurs and as many Popes. Methinks I have outlived myself, and begin to be weary of the sun. I have shaken hands with delight in my warm blood and canicular days. I perceive I do anticipate the vices of age: the world to me is but a dream or mock-show, and we all therein but pantaloons and antics to my severer contemplations. The course and order of my life would be a very death unto another. I use myself to all diets, humours, airs, hunger, thirst, cold, heat, want, plenty, necessity, dangers, hazards. When I am cold I cure not myself by heat; when sick, not by physic. Those that know how I live may justly say I regard not life, nor stand in fear of death.

Section 42. It is not, I confess, an unlawful prayer to desire to surpass the days of our Saviour, or wish to outlive that age wherein he thought fittest to die, yet if—as divinity affirms—there shall be no grey hairs in heaven, but all shall rise in the perfect state of men, we do but outlive those perfections in this world to be recalled unto them by a greater

miracle in the next, and run on here but to be retrograde hereafter. Were there any hopes to outlive vice, or a point to be
superannuated from sin, it were worthy of our knees to implore the days of Methuselah. But age doth not rectify but
incurvate our natures, turning bad dispositions into worser
habits, and—like diseases—brings on incurable vices; for every
day as we grow weaker in age we grow stronger in sin, and
the number of our days doth but make our sins innumerable.
The same vice committed at sixteen is not the same, though it
agree in all other circumstances, at forty, but swells and 10
doubles from the circumstances of our ages, when besides the
constant and inexcusable habit of transgressing, the maturity
of our judgement cuts off pretence unto excuse or pardon.
Every sin, the oftener it is committed, the more it acquireth
in the quality of evil: as it succeeds in time, so it proceeds in
degrees of badness; for as they proceed they ever multiply,
and, like figures in arithmetic, the last stands for more than all
that went before it. And though I think no man can live well
once but he that could live twice, yet for my own part
I would not live over my hours past, or begin again the thread 20
of my days: not upon Cicero's ground, because I have lived
them well, but for fear I should live them worse. I find my
growing judgement daily instruct me how to be better, but
my untamed affections and confirmed vitiosity makes me
daily do worse. I find in my confirmed age the same sins
I discovered in my youth: I committed many then because
I was a child, and because I commit them still, I am yet an
infant. Therefore I perceive a man may be twice a child before
the days of dotage, and stand in need of Aeson's bath before
threescore.   30

Section 43. And truly there goes a great deal of providence
to produce a man's life unto threescore. There is more required
than an able temper for those years. Though the radical

humour contain in it sufficient oil for seventy, yet I perceive in some it gives no light past thirty; men assign not all the causes of long life that write whole books thereof. They that found themselves on the radical balsam or vital sulphur of the parts determine not why Abel lived not so long as Adam. There is therefore a secret gloom or bottom of our days; 'twas his wisdom to determine them, but his perpetual and waking providence that fulfils and accomplisheth them; wherein the spirits, ourselves and all the creatures of God in a secret and
10 disputed way do execute his will. Let them not therefore complain of immaturity that die about thirty. They fall but like the whole world, whose solid and well-composed substance must not expect the duration and period of its constitution. When all things are completed in it, its age is accomplished; and the last and general fever may as naturally destroy it before six thousand as me before forty. There is therefore some other hand that twines the thread of life than that of nature. We are not only ignorant in antipathies and occult qualities; our ends are as obscure as our beginnings. The line
20 of our days is drawn by night, and the various effects therein by a pencil that is invisible; wherein though we confess our ignorance, I am sure we do not err if we say it is the hand of God.

Section 44. I am much taken with two verses of Lucan since I have been able not only—as we do at school—to construe, but understand them:

> *Victurosque dei celant, ut vivere durent,*
> *Felix esse mori.*
> We're all deluded, vainly searching ways
> To make us happy by the length of days;
30   For cunningly to make's protract this breath,
> The gods conceal the happiness of death.

There be many excellent strains in that poet, wherewith his stoical genius hath liberally supplied him; and truly there are singular pieces in the philosophy of Zeno and doctrine of the Stoics which I perceive, delivered in a pulpit, pass for current divinity. Yet herein are they in extremes, that can allow a man to be his own assassin, and so highly extol the end and suicide of Cato. This is indeed not to fear death, but yet to be afraid of life. It is a brave act of valour to contemn death; but where life is more terrible than death it is then the truest valour to dare to live, and herein religion hath taught us a noble example. For all the valiant acts of Curtius, Scaevola or Codrus do not parallel or match that one of Job; and sure there is no torture to the rack of a disease, nor any poniards in death itself like those in the way and prologue unto it. *Emori nolo, sed me esse mortuum nihil curo*; I would not die, but care not to be dead. Were I of Caesar's religion I should be of his desires, and wish rather to go off at one blow than to be sawed in pieces by the grating torture of a disease. Men that look no further than their outsides think health an appurtenance unto life, and quarrel with their constitutions for being sick; but I that have examined the parts of man, and know upon what tender filaments that fabric hangs, do wonder that we are not always so; and considering the thousand doors that lead to death, do thank my God that we can die but once. 'Tis not only the mischief of diseases and the villainy of poisons that make an end of us. We vainly accuse the fury of guns and the new inventions of death: 'tis in the power of every hand to destroy us, and we are beholding unto everyone we meet he doth not kill us. There is therefore but one comfort left; that though it be in the power of the weakest arm to take away life, it is not in the strongest to deprive us of death. God would not exempt himself from that; the misery of immortality in the flesh he undertook not, that was in it immortal. Certainly there is no happiness within this circle of flesh, nor is it in the optics of these eyes to behold felicity: the first day of our jubilee is death. The devil hath therefore failed of his desires: we are

53

happier with death than we should have been without it. There is no misery but in himself, where there is no end of misery; and so indeed, in his own sense, the Stoic is in the right. He forgets that he can die who complains of misery. We are in the power of no calamity while death is in our own.

Section 45. Now besides this literal and positive kind of death, there are others whereof divines make mention, and those, I think, not merely metaphorical; as mortification, dying unto sin and the world. Therefore I say every man hath a double horoscope; one of his humanity, his birth; another of his Christianity, his baptism; and from this do I compute and calculate my nativity, not reckoning those *horae combustae* and odd days, or esteeming myself anything before I was my Saviour's, and enrolled in the register of Christ. Whosoever enjoys not this life, I count him but an apparition, though he wear about him the sensible affections of flesh. In these moral acceptions the way to be immortal is to die daily, nor can I think I have the true theory of death when I contemplate a skull or behold a skeleton with those vulgar imaginations it casts upon us. I have therefore enlarged that common *memento mori* into a more Christian memorandum, *Memento quatuor novissima*, those four inevitable points of us all, death, judgement, heaven and hell. Neither did the contemplations of the heathens rest in their graves without a further thought of Rhadamanth, or some judicial proceeding after death, though in another way and upon suggestion of their natural reasons. I cannot but marvel from what Sibyl or oracle they stole the prophecy of the world's destruction by fire, or whence Lucan learned to say,

> *Communis mundo superest rogus, ossibus astra*
> *Misturus.*
> There yet remains to th'world one common fire,
> Wherein our bones with stars shall make one pyre.

54

I do believe the world draws near its end, yet is neither old nor decayed, nor will ever perish upon the ruins of its own principles. As the Creation was a work above nature, so is its adversary, annihilation; without which the world hath not its end but its mutation. Now what fire should be able to consume it thus far without the breath of God, which is the truest consuming flame, my philosophy cannot inform me. Some believe there went not a minute to the world's creation, nor shall there go to its destruction. Those six days so punctually described make not to them one moment, but rather seem to manifest the method and idea of the great work of the intellect of God than the manner how he proceeded in its operation. I cannot dream there should be at the last day any such judicial proceeding or calling to the bar as indeed the Scripture seems to imply, and the literal commentators do conceive; for unspeakable mysteries in the Scriptures are often delivered in a vulgar and illustrative way, and being written unto man are delivered not as they truly are, but as they may be understood; wherein, notwithstanding, the different interpretations according to different capacities, may stand firm with our devotion, nor be any way prejudicial to each single edification.

Section 46. Now to determine the day and year of this inevitable time is not only convincible and statute madness, but also manifest impiety. How shall we interpret Elias' six thousand years, or imagine the secret communicated to a rabbi which God hath denied to his angels? It had been an excellent query to have posed the devil of Delphos, and must needs have forced him to some strange amphibology. It hath not only mocked the predictions of sundry astrologers in ages past, but the prophecies of many melancholy heads in these present, who neither reasonably understanding things past or present, pretend a knowledge of things to come; heads ordained only to manifest the incredible effects of melancholy, and to fulfil old prophecies rather than be the authors of new. Those prognostics of Scripture are obscure; I know not how to construe them. 'In those days there shall come wars and rumours of wars', to me seems no prophecy but a constant truth, in all times verified since it was first pronounced. 'There shall be signs in the moon and stars': how comes he then like a thief in the night, when he gives an item of his coming? That common sign drawn from the revelation of Antichrist is as obscure as any. In our common compute he hath been 10 come these many years. But for my own part, to speak freely, omitting those ridiculous anagrams, I am half of Paracelsus' opinion, and think Antichrist the philosopher's stone in divinity, for the discovery and invention whereof though there be prescribed rules and probable inductions, yet hath hardly any man attained the perfect discovery thereof. That general opinion that the world draws near its end hath possessed all ages past as nearly as ours. I am afraid that the souls that now depart cannot escape that lingering expostulation of the saints under the altar, *Quousque, Domine?* How 20 long, O Lord? and groan in the expectation of that great jubilee.

Section 47. This is the day that must make good that great attribute of God, his justice; that must reconcile those unanswerable doubts which torment the wisest understandings, and reduce those seeming inequalities and respective distributions in this world to an equality and recompensive justice in the next. This is that one day that shall include and comprehend all that went before it, wherein as in the last scene all the actors must enter, to complete and make up the catastrophe 30 of this great piece. This is the day whose memory hath only power to make us honest in the dark, and to be virtuous without a witness. *Ipsa sui pretium virtus sibi*; that virtue is her own reward is but a cold principle, and not able to maintain our variable resolutions in a constant and settled way of goodness. I have practised that honest artifice of Seneca, and in my retired and solitary imaginations, to detain me from the foulness of vice, have fancied to myself the presence of my dear and worthiest friends, before whom I should lose my head rather than be vicious. Yet herein I found that there was naught but moral honesty, and this was not to be virtuous for his sake that must reward us at the last. I have tried if I could reach that great resolution of his and be honest without a thought of heaven or hell; and indeed I found upon a natural inclination and inbred loyalty unto virtue that I could serve her without a livery, yet not in that resolved and venerable way but that the frailty of my nature, upon an easy temptation, might be induced to forget her. The life therefore and spirit of all our actions is the Resurrection, and stable apprehension that our ashes shall enjoy the fruits of our pious endeavours. Without this, all religion is a fallacy, and those impieties of Lucian, Euripides and Julian are no blasphemies but subtle verities, and atheists have been the only philosophers.

Section 48. How shall the dead arise is no question of my faith. To believe only possibilities is not faith but mere philosophy. Many things are true in divinity which are neither inducible by reason nor confirmable by sense, and many things in philosophy confirmable by sense yet not inducible by reason. Thus it is impossible by any solid or demonstrative reasons to persuade a man to believe the conversion of the needle to the north; though this be possible, and true and easily credible, upon a single experiment unto the sense. I believe that our estranged and divided ashes shall unite again; that our separated dust, after so many pilgrimages and transformations into the parts of minerals, plants, animals, elements, shall at the voice of God return into their primitive

shapes, and join again to make up their primary and predestinated forms. As at the Creation of the world there was
a separation of that confused mass into its species, so at the
destruction thereof there shall be a separation into its distinct
individuals. As at the Creation of the world all the distinct
species that we behold lay involved in one mass, till the
fruitful voice of God separated this united multitude into its
several species, so at the last day, when those corrupted relics
shall be scattered in the wilderness of forms and seem to have
10 forgot their proper habits, God by a powerful voice shall
command them back into their proper shapes, and call them
out by their single individuals. Then shall appear the fertility
of Adam, and the magic of that sperm that hath dilated into so
many millions. What is made to be immortal, nature cannot—
nor will the voice of God—destroy. These bodies that we
behold to perish were in their created natures immortal, and
liable unto death but accidentally and upon forfeit; and therefore they owe not that natural homage unto death as other
bodies, but may be restored to immortality with a lesser
20 miracle, and by a bare and easy revocation of the curse return
immortal. I have often beheld—like a miracle—that artificial
resurrection and revivification of mercury; how, being mortified into a thousand shapes it assumes again its own, and
returns into its numerical self. Let us speak naturally and like
philosophers, the forms of alterable bodies in their sensible
corruption perish not, nor, as we imagine, wholly quit their
mansions, but retire and contract themselves into their secret
and inaccessible parts, where they may best protect themselves
from the action of their antagonist. A plant or vegetable con-
30 sumed to ashes, to a contemplative and School philosopher
seems utterly destroyed, and the form to have taken his leave
for ever: but to a sensible artist the forms are not perished
but withdrawn into their incombustible part, where they lie
secure from the action of that devouring element. This is made

good by experience, which can from the ashes of a plant revivify the plant, and from its cinders recall it into its stalk and leaves again. What the art of man can do in these inferior pieces, what blasphemy is it to affirm the finger of God cannot do in these more perfect and sensible structures! This is that mystical philosophy from whence no true scholar becomes an atheist, but from the visible effects of nature grows up a real divine, and beholds not in a dream, as Ezekiel, but in an ocular and visible object the types of his resurrection.

Section 49. Now the necessary mansions of our restored selves are those two contrary and incompatible places we call heaven and hell; for to define them or strictly to determine what and where they are surpasseth my divinity. That elegant apostle which seemed to have a glimpse of heaven hath left but a negative description thereof; which neither eye hath seen, nor ear hath heard, nor can it enter into the heart of man. He was translated out of himself to behold it, but being returned into himself, could not express it. St John's description by emeralds, chrysolites and precious stones is too weak to express the material heaven we behold. Briefly therefore, where the soul hath the full measure and complement of happiness, where the boundless appetite of that spirit remains completely satisfied, that it cannot desire either addition or alteration, that I think is truly heaven; and this can only be in the enjoyment of that essence whose infinite goodness is able to terminate the desires of itself and the insatiable wishes of ours. Wherever God will thus manifest himself, there is heaven, though within the circle of this sensible world. Thus the soul of man may be in heaven anywhere, even within the limits of his own proper body, and when it ceaseth to live in the body it may remain in its own soul, that is, its Creator: and thus we may say that St Paul, whether in the body or out of the body, was yet in heaven. To place it in the empyreal, or

beyond the tenth sphere, is to forget the world's destruction; for when this sensible world shall be destroyed, all shall then be here as it is now there; an empyreal heaven, a *quasi* vacuity, or place exempt from the natural affection of bodies, where to ask where heaven is, is to demand where the presence of God is, or where we have the glory of that happy vision. Moses, that was bred up in all the learning of the Egyptians, committed a gross absurdity in philosophy when with these eyes of flesh he desired to see God, and petitioned his Maker, that
10 is truth itself, to a contradiction. Those that imagine heaven and hell neighbours, and conceive a vicinity between these two extremes upon consequence of the parable where Dives discoursed with Lazarus in Abraham's bosom, do too grossly conceive of those glorified creatures whose eyes shall easily out-see the sun and behold without a perspective the extremest distances. For if there shall be in our glorified eyes the faculty of sight and reception of objects, I could think the visible species there to be in as unlimitable a way as now the intellectual. I grant that two bodies placed beyond the tenth sphere
20 or in a vacuity, according to Aristotle's philosophy, could not behold each other, because there wants a body or medium to hand and transport the visible rays of the object unto the sense; but when there shall be a general defect of either medium to convey or light to prepare and dispose that medium, and yet a perfect vision, we must suspend the rules of our philosophy, and make all good by a more absolute piece of optics.

Section 50. I cannot tell how to say that fire is the essence of hell: I know not what to make of purgatory, or conceive a
30 flame that can either prey upon or purify the substance of a soul. Those flames of sulphur mentioned in the Scripture I take to be understood not of this present hell, but of that to come, where fire shall make up the complement of our tortures,

and have a body or subject whereon to manifest its tyranny.
Some who have had the honour to be textuary in divinity are
of opinion it shall be the same specifical fire with ours. This
is hard to conceive; yet can I make good how even that may
prey upon our bodies, and yet not consume us: for in this
material world there are bodies that persist invincible in the
powerfullest flames; and though by the action of fire they fall
into ignition and liquation, yet will they never suffer a
destruction. I would gladly know how Moses with an actual
fire calcined or burnt the golden calf into powder; for that 10
mystical metal of gold, whose solary and celestial nature
I admire, exposed unto the violence of fire grows only hot and
liquefies, but consumeth not, neither in its substance, weight or
virtue. So, when the consumable and volatile pieces of our
bodies shall be refined into a more impregnable and fixed
temper like gold, though they suffer from the action of flames
they shall never perish, but lie immortal in the arms of fire.
And surely, if this frame must suffer only by the action of this
element, there will many bodies escape; and not only heaven
but earth will not be at an end, but rather a beginning. For at 20
present it is not earth but a composition of fire, water, earth
and air: but at that time, spoiled of these ingredients, it shall
appear in a substance more like itself, its ashes. Philosophers
that opinioned the world's destruction by fire did never dream
of annihilation, which is beyond the power of sublunary
causes; for the last and powerfullest action of that element is
but vitrification, or a reduction of a body into glass; and therefore some of our chemists facetiously affirm—yea, and urge
Scripture for it—that at the last fire all shall be crystallized and
reverberated into glass, which is the utmost action of that 30
element. Nor need we fear this term annihilation, or wonder
that God will destroy the works of his creation; for man
subsisting—who is and will then truly appear a microcosm—
the world cannot be said to be destroyed. For the eyes of

God, and perhaps also of our glorified senses, shall as really
behold and contemplate the world in its epitome or con-
tracted essence as now they do at large and in its dilated substance. In the seed of a plant, to the eyes of God and to the
understanding of man there exists, though in an invisible way,
the perfect leaves, fruits and flowers thereof: for things that
are *in posse* to the sense are actually existent to the understanding. Thus God beholds all things, who contemplates as
fully his works in their epitome as in their full volume, and
10 beheld as amply the whole world in that little compendium
of the sixth day as in the scattered and dilated pieces of those
five before.

Section 51. Men commonly set forth the tortures of hell by
fire and the extremity of corporal afflictions, and describe hell
in the same manner as Mahomet doth heaven. This indeed
makes a noise and drums in popular ears; but if this be the
terrible piece thereof it is not worthy to stand in diameter
with heaven, whose happiness consists in that part which is
best able to comprehend it—that immortal essence, that trans-
20 lated divinity and colony of God, the soul. Surely, though we
place hell under earth, the devil's walk and purlieu is about it.
Men speak too popularly who place it in those flaming mountains which to grosser apprehensions represent hell. The heart
of man is the place the devil dwells in. I feel sometimes a hell
within myself; Lucifer keeps his court in my breast, Legion is
revived in me. There are as many hells as Anaxagoras conceited worlds. There was more than one hell in Magdalene
when there were seven devils, for every devil is an hell unto
himself. He holds enough of torture in his own *ubi*, and
30 needs not the misery of circumference to afflict him; and thus
a distracted conscience here is a shadow or introduction unto
hell hereafter. Who can but pity the merciful intention of
those hands that do destroy themselves? The devil, were it in

62

his power, would do the like; which being impossible, his miseries are endless, and he suffers most in that attribute wherein he is impassible, his immortality.

Section 52. I thank God—and with joy I mention it—I was never afraid of hell, nor never grew pale at the description of that place. I have so fixed my contemplations on heaven that I have almost forgot the idea of hell: I am afraid rather to lose the joys of the one than endure the misery of the other: to be deprived of them is a perfect hell, and needs, methinks, no addition to complete our afflictions. That terrible term hath never detained me from sin, nor do I owe one good action to the name thereof. I fear God, yet am not afraid of him; his mercies make me ashamed of my sins before his judgements afraid thereof. These are the forced and secondary method of his wisdom, which he useth not but as the last remedy and upon provocation; a course rather to deter the wicked than incite the virtuous to his worship. I can hardly think there was ever any scared into heaven. They go the surest way to heaven who would serve God without a hell. Other mercenaries, that crouch unto him in fear of hell, though they term themselves the servants are indeed but the slaves of the Almighty.

Section 53. And to be true and speak my soul, when I survey the occurrences of my life and call into account the finger of God, I can perceive nothing but an abyss and mass of his mercies, either in general to mankind or in particular to myself. And whether out of the prejudice of my affection or an inverting and partial conceit of his mercies I know not, but those which others term crosses, afflictions, judgements, misfortunes, to me, who enquire further into them than their visible effects, they both appear and in event have ever proved the secret and dissembled favours of his affection. It is a singular piece of wisdom to apprehend truly and without passion the works of God, and so well to distinguish his justice from his mercy as not to miscall those noble attributes; yet it is likewise an honest piece of logic so to dispute and argue the proceedings of God as to distinguish even his judgements into mercies. For God is merciful unto all, because better to the worst than the best deserve; and to say he punisheth none in this world, though it be a paradox, is no absurdity. To one that hath committed murder, if the judge should only ordain a box of the ear, it were a madness to call this a punishment, and to repine at the sentence rather than admire the clemency of the judge. Thus our offences being mortal, and deserving not only death but damnation, if the goodness of God be content to traverse and pass them over with a loss, misfortune, or disease, what frenzy were it to term this a punishment rather than an extremity of mercy, and to groan under the rod of his judgements rather than admire the sceptre of his mercies? Therefore to adore, honour and admire him is a debt of gratitude due from the obligation of our natures, states, and conditions; and with these thoughts, he that knows them best will not deny that I adore him. That I obtain heaven or the bliss thereof is accidental, and not the intended work of my devotion, it being a felicity I can neither think to deserve nor scarce in modesty expect. For these two ends of us all, either as rewards or punishments, are mercifully ordained and disproportionally disposed unto our actions; the one being so far beyond our deserts, the other so infinitely below our demerits.

Section 54. There is no salvation to those that believe not in Christ—that is, say some, since his Nativity, and as divinity affirmeth, before also; which makes me much apprehend the end of those honest worthies and philosophers which died before his Incarnation. It is hard to place those souls in hell whose worthy lives do teach us virtue on earth. Methinks amongst those many subdivisions of hell there might have been one limbo left for these. What a strange vision will it be to see their poetical fictions converted into verities, and their imaginary and fancied furies into real devils! How strange to them will sound the history of Adam, when they shall suffer for him they never heard of! when they that derive their genealogy from the gods shall know they are the unhappy issue of sinful man! It is an insolent part of reason to controvert the works of God, or question the justice of his proceedings. Could humility teach others, as it hath instructed me, to contemplate the infinite and incomprehensible distance betwixt the Creator and the creature, or did we seriously perpend that one simile of St Paul, *Shall the vessel say unto the potter, Why hast thou made me thus?* it would prevent these arrogant disputes of reason; nor would we argue the definitive sentence of God, either to heaven or hell. Men that live according to the right rule and law of reason live but in their own kind, as beasts do in theirs; who justly obey the prescript of their natures, and therefore cannot reasonably demand a reward of their actions, as only obeying the natural dictates of their reason. It will, therefore, and must at last appear that all salvation is through Christ; which verily I fear these great examples of virtue must confirm, and make it good how the perfectest actions of earth have no title or claim unto heaven.

Section 55. Nor truly do I think the lives of these or any other were ever correspondent or in all points conformable unto their doctrines. It is evident that Aristotle transgressed the rule of his own ethics. The Stoics, that condemn all passion and command a man to laugh in Phalaris' bull, could not endure without a groan a fit of the stone or colic. The Sceptics, that affirmed they knew nothing, even in that opinion confuted themselves, and thought they knew more than all the world beside. Diogenes I hold to be the most vainglorious man of his time, and more ambitious in refusing all honours than Alexander in rejecting none. Vice and the devil put a fallacy upon our reasons, and provoking us too hastily to run from it, entangle and profound us deeper in it. The Duke of Venice, that yearly weds himself unto his sea by casting therein a ring of gold, I will not argue of prodigality, because it is a solemnity of good use and consequence in the state. But the philosopher that threw his money into the sea to avoid avarice was a notorious prodigal. There is no road or ready way to virtue. It is not an easy point of art to untangle ourselves from this riddle and web of sin. To perfect virtue, as to religion, there is required a panoplia or complete armour, that whilst we lie at close ward against one vice, we lie not open to the venue of another. And indeed, wiser discretions, that have the thread of reason to conduct them, offend without a pardon, whereas under-heads may stumble without dishonour. There go so many circumstances to piece up one good action that it is a lesson to be good, and we are forced to be virtuous by the book. Again, the practice of men holds not an equal pace—yea, and often runs counter to their theory. We naturally know what is good, but naturally pursue what is evil. The rhetoric wherewith I persuade another cannot persuade myself: there is a depraved appetite in us that will with patience hear the learned instructions of reason, but yet perform no farther than agrees to its own irregular humour. In brief, we are all monsters—that is, a composition of man and beast; wherein we must endeavour to be as the poets feign that wise man Chiron; that is, to have the region of man above that of beast, and sense to sit but at the feet of reason. Lastly, I do desire with God that all—but yet affirm with men that very few—shall know salvation; that the bridge is narrow, the passage strait unto life; yet those who do confine the Church of God either to particular nations, churches, or families have made it far narrower than our Saviour ever meant it.

Section 56. The vulgarity of those judgements that wrap the Church of God in Strabo's cloak and restrain it unto Europe seem to me as bad geographers as Alexander, who thought he had conquered all the world when he had not subdued the half of any part thereof. For we cannot deny the Church of God both in Asia and Africa if we do not forget the peregrinations of the Apostles, the death of their martyrs, the 10 sessions of many, and—even in our reformed judgement— lawful councils held in those parts in the minority and nonage of ours. Nor must a few differences, more remarkable in the eyes of man than perhaps in the judgement of God, excommunicate from heaven one another, much less those Christians who are in a manner all martyrs, maintaining their faith in the noble way of persecution and serving God in the fire, whereas we honour him but in the sunshine. 'Tis true we all hold there is a number of elect and many to be saved, yet take our opinions together and from the confusion thereof there will 20 be no such thing as salvation, nor shall anyone be saved. For first the Church of Rome condemneth us, we likewise them; the Sub-reformists and Sectaries sentence the doctrine of our Church as damnable, the Atomist or Familist reprobates all these, and all these them again. Thus whilst the mercies of God do promise us heaven, our conceits and opinions exclude us from that place. There must be therefore more than one St Peter; particular churches and sects usurp the gates of heaven and turn the key against each other; and thus we go to heaven against each other's wills, conceits and opinions, 30 and with as much uncharity as ignorance do err, I fear, in points not only of our own but one another's salvation.

Section 57. I believe many are saved who to man seem reprobated, and many reprobated who in the opinion and sentence of man stand elected. There will appear at the Last Day strange and unexpected examples both of his justice and his mercy, and therefore to define either is folly in man, and insolency even in the devils. Those acute and subtle spirits, in all their sagacity, can hardly divine who shall be saved; which if they could prognostic their labour were at an end, nor need they compass the earth seeking whom they may devour.
10 Those who upon a rigid application of the law sentence Solomon unto damnation condemn not only him but themselves and the whole world; for by the letter and written law of God we are without exception in the state of death. But there is a prerogative of God, and an arbitrary pleasure above the letter of his own law, by which alone we can pretend unto salvation, and through which Solomon might as easily be saved as those who condemn him.

Section 58. The number of those who pretend unto salvation, and those infinite swarms who think to pass through the
20 eye of this needle, hath much amazed me. That name and compellation of 'little flock' doth not comfort but deject my devotion, especially when I reflect upon mine own unworthiness, wherein—according to my humble apprehensions—I am below them all. I believe there shall never be an anarchy in heaven; but as there are hierarchies amongst the angels, so shall there be degrees of priority amongst the saints. Yet is it, I protest, beyond my ambition to aspire unto the first ranks. My desires only are—and I shall be happy therein— to be but the last man, and bring up the rear in heaven.

30 *Section 59.* Again, I am confident and fully persuaded, yet dare not take my oath of my salvation. I am as it were sure, and do believe without all doubt that there is such a city as

68

Constantinople, yet for me to take my oath thereon it were a kind of perjury, because I hold no infallible warrant from my own sense to confirm me in the certainty thereof. And truly, though many pretend an absolute certainty of their salvation, yet when an humble soul shall contemplate her own unworthiness she shall meet with many doubts, and suddenly find how little we stand in need of the precept of St Paul, *Work out your salvation with fear and trembling*. That which is the cause of my election I hold to be the cause of my salvation, which was the mercy and beneplacit of God, before I was, or the foundation of the world. *Before Abraham was, I am*, is the saying of Christ; yet is it true in some sense if I say it of myself, for I was not only before myself but Adam; that is, in the idea of God and the decree of that synod held from all eternity. And in this sense, I say, the world was before the Creation, and at an end before it had a beginning; and thus was I dead before I was alive. Though my grave be England my dying place was Paradise, and Eve miscarried of me before she conceived of Cain.

Section 60. Insolent zeals, that do decry good works and rely only upon faith, take not away merits; for depending upon the efficacy of their faith they enforce the condition of God, and in a more sophistical way do seem to challenge heaven. It was decreed by God that only those that lapped in the waters like dogs should have the honour to destroy the Midianites, yet could none of these challenge or imagine he deserved that honour thereupon. I do not deny but that true faith, and such as God requires, is not only a mark or token but also a means of our salvation; but where to find this is as obscure to me as my last end. And if our Saviour could object unto his own disciples and favourites a faith that to the quantity of a grain of mustard-seed is able to remove mountains, surely that which we boast of is not anything, or at the most but a remove from nothing. This is the tenor of my belief, wherein though there be many things singular and to the humour of my irregular self, yet if they square not with maturer judgements I disclaim them, and do no further father them than the learned and best judgements shall authorise them.

Section 1. Now for that other virtue of charity, without which faith is a mere notion and of no existence, I have ever endeavoured to nourish the merciful disposition and humane inclination I borrowed from my parents, and to regulate it to the written and prescribed laws of charity. And if I hold the true anatomy of myself, I am delineated and naturally framed to such a piece of virtue, for I am of a constitution so general that it consorts and sympathiseth with all things. I have no antipathy—or rather, idiosyncrasy—in diet, humour, air, any-thing. I wonder not at the French for their dishes of frogs, 10 snails and toadstools, nor at the Jews for locusts and grass-hoppers, but being amongst them make them my common viands; and I find they agree with my stomach as well as theirs. I could digest a salad gathered in a churchyard as well as in a garden. I cannot start at the presence of a serpent, scorpion, lizard or salamander; at the sight of a toad or viper I feel in me no desire to take up a stone to destroy them. I find not in myself those common antipathies that I can discover in others. Those national repugnances do not touch me, nor do I behold with prejudice the French, Italian, Spaniard or 20 Dutch; but where I find their actions in balance with my countrymen's I honour, love and embrace them in the same degree. I was born in the eighth climate, but seem to be framed and constellated unto all: I am no plant that will not prosper out of a garden. All places, all airs, make unto me one country; I am in England everywhere, and under any meridian. I have been shipwrecked, yet am not enemy with the sea or winds; I can study, play or sleep in a tempest. In brief, I am averse from nothing, neither plant, animal nor spirit. My conscience would give me the lie if I should say 30 I absolutely detest and hate any essence but the devil, or so at

least abhor anything but that we might come to composition.
If there be any among those common objects of hatred which
I can safely say I do contemn and laugh at, it is that great
enemy of reason, virtue and religion, the multitude; that
numerous piece of monstrosity which taken asunder seem men
and the reasonable creatures of God, but confused together
make but one great beast, and a monstrosity more prodigious
than Hydra. It is no breach of charity to call these fools; it is
the style all holy writers have afforded them, set down by
10 Solomon in canonical Scripture, and a point of our faith to
believe so. Neither in the name of multitude do I only include
the base and minor sort of people. There is a rabble even
amongst the gentry, a sort of plebeian heads whose fancy
moves with the same wheel as these; men in the same level
with mechanics, though their fortunes do somewhat gild
their infirmities, and their purses compound for their follies.
But as in casting of account three or four men together come
short of one man placed by himself below them, so neither
are a troop of these ignorant doradoes of that true esteem
20 and value as many a forlorn person whose condition doth
place them below their feet. Let us speak like politicians,
there is a nobility without heraldry, a natural dignity whereby
one man is ranked with another, another filed before him,
according to the quality of his desert and pre-eminence of his
good parts. Though the corruption of these times and the bias
of present practice wheel another way, thus it was in the first
and primitive commonwealths, and is yet in the integrity and
cradle of well-ordered polities till corruption getteth ground;
ruder desires labouring after that which wiser considerations
30 contemn, everyone having a liberty to amass and heap up
riches, and they a licence or faculty to do or purchase anything.

Section 2. This general and indifferent temper of mine doth
more nearly dispose me to this noble virtue, that with an

easier measure of grace I may obtain it. It is a happiness to
be born and framed unto virtue, and to grow up from the
seeds of nature rather than the inoculation and forced grafts
of education; yet if we are directed only by our particular
natures, and regulate our inclinations by no higher rule than
that of our reasons, we are but moralists; divinity will still
call us heathens. Therefore this great work of charity must
have other motives, ends and impulsives. I give no alms only
to satisfy the hunger of my brother, but to fulfil and accom-
plish the will and command of my God. I draw not my purse 10
for his sake that demands it, but his that enjoined it. I relieve
no man upon the rhetoric of his miseries, nor to content mine
own commiserating disposition; for this is still but moral
charity, and an act that oweth more to passion than reason.
He that relieves another upon the bare suggestion and bowels
of pity doth not this so much for his sake as for his own; for
by compassion we make another's misery our own, and so by
relieving them we relieve ourselves also. It is as erroneous
a conceit to redress other men's misfortunes upon that common consideration of merciful natures that it may be one day 20
our own case; for this is a sinister and politic kind of charity
whereby we seem to bespeak the pities of men in the like
occasions—buy out of God a faculty to be exempted from it.
And truly, I have observed that these professed eleemosynaries,
though in a crowd or multitude, do yet direct and place their
petitions on a few and selected persons. There is surely a
physiognomy which those experienced and master-mendicants
observe, whereby they instantly discover a merciful aspect,
and will single out a face wherein they spy the signatures and
marks of mercy: for there are mystically in our faces certain 30
characters which carry in them the motto of our souls, wherein
he that cannot read a, b, c, may read our natures. I hold
moreover that there is a phytognomy or physiognomy not
only of men but of plants and vegetables, and in every one of

them some outward figures which hang as signs and bushes of their inward forms. The finger of God hath set an inscription upon all his works, not graphical or composed of letters, but of their several forms, constitutions, parts and operations, which aptly joined together make one word that doth express their natures. By these letters God calls the stars by their names, and by this alphabet Adam assigned to every creature a name peculiar to its nature. Now there are, besides these characters in our faces, certain mystical lines and figures in our hands
10 which I dare not call mere dashes or strokes *à la volée*, or at random, because delineated by a pencil that never works in vain; and hereof I take more particular notice because I carry that in mine own hand which I could never read of or discover in another. Aristotle, I confess, in his acute and singular book of physiognomy hath made no mention of chiromancy, yet I believe the Egyptians, who were nearer addicted to those abstruse and mystical sciences, had a knowledge therein to which those vagabond and counterfeit Egyptians did after pretend, and perhaps retained a few corrupted principles
20 which sometimes might verify their prognostics.

It is the common wonder of all men how among so many millions of faces there should be none alike. Now contrary, I wonder as much how there should be any. He that shall consider how many thousand several words have been carelessly and without study composed out of twenty-four letters, withal how many hundred lines there are to be drawn in the fabric of one man, shall easily find that this variety is necessary; and it will be very hard that they should so concur as to make one portrait like another. Let a painter carelessly limn out a
30 million of faces, and you shall find them all different; yea, let him have his copy before him, yet after all his art there will remain a sensible distinction: for the pattern or example of everything is the perfectest in that kind, whereof we still come short though we transcend and go beyond it, because

herein it is wide and agrees not in all points unto its copy. I rather wonder how almost all plants, being of one colour, yet should be all different herein and their several kinds distinguished in one accident of vert. Nor doth the similitude of creatures disparage the variety of nature, nor any way confound the works of God. For even in things alike there is a diversity, and those that do seem to accord do manifestly disagree. And thus is man like God, for in the same things that we resemble him, we are utterly different from him. There is never anything so like another as in all points to 10 concur; there will ever some reserved difference slip in to prevent the identity, without which two several things would not be alike but the same, which is impossible.

Section 3. But to return from philosophy to charity, I hold not so narrow a conceit of this virtue as to conceive to give alms is only to be charitable, or think a piece of liberality can comprehend the total of charity. Divinity hath wisely divided the acts thereof into many branches, and hath taught us in this narrow way many paths unto goodness. As many ways as we may do good, so many ways we may be charitable. 20 There are infirmities not only of body but of soul and fortunes, which do require the merciful hand of our abilities. I cannot contemn a man for ignorance, but behold him with as much pity as I do Lazarus. It is no greater charity to clothe his body than apparel the nakedness of his soul. It is an honourable object to see the reasons of other men wear our liveries, and their borrowed understandings do homage to the bounty of ours. It is the cheapest way of beneficence, and like the natural charity of the sun, illuminates another without obscuring itself. To be reserved and caitiff in this part of good- 30 ness is the sordidest part of covetousness, and more contemptible than pecuniary avarice. To this—as calling myself a scholar—I am obliged by the duty of my condition: I make

75

not therefore my head a grave but a treasure of knowledge.
I intend no monopoly, but a community in learning: I study
not for my own sake only, but for theirs that study not for
themselves. I envy no man that knows more than myself, but
pity those that know less. I instruct no man as an exercise of
my knowledge, or with intent rather to nourish and keep it
alive in mine own head than beget and propagate it in his; and
in the midst of all my endeavours there is but one thought
that dejects me—that my acquired parts must perish with
10 myself, nor can be legacied among my honoured friends.
I cannot fall out or contemn a man for an error, or conceive
why a difference in opinion should divide an affection: for
controversies, disputes and argumentations, both in philosophy and divinity, if they meet with discreet and peaceable
natures, do not infringe the laws of charity. In all disputes,
so much as there is of passion so much there is of nothing to
the purpose; for then reason, like a bad hound, spends upon
a false scent and forsakes the question first started. And this is
one reason why controversies are never determined; for
20 though they be amply proposed they are scarce at all handled,
they do so swell with unnecessary digressions; and the
parenthesis on the party is often as large as the main discourse
upon the subject. The foundations of religion are already
established and the principles of salvation subscribed unto by
all. There remain not many controversies worth a passion, and
yet never any disputed without, not only in divinity but
inferior arts. What a Βατραχομυομαχία and hot skirmish is
betwixt σ and τ in Lucian! How do the grammarians hack
and slash for the genitive case in Jupiter! How many synods
30 have been assembled and angerly broke up about a line in
Propria quae maribus! How do they break their own pates
to save that of Priscian! *Si foret in terris, rideret Democritus.*
Yea, even amongst wiser militants, how many wounds have
been given and credits slain for the poor victory of an opinion,

or beggarly conquest of a distinction! Scholars are men of peace, they bear no arms; but their tongues are sharper than Actius' razor, their pens carry farther and give a louder report than thunder. I had rather stand the shock of a basilisco than the fury of a merciless pen. It is not mere zeal to learning or devotion to the Muses that wiser princes patron the arts and carry an indulgent aspect unto scholars, but a desire to have their names eternised by the memory of their writings, and a fear of the revengeful pen of succeeding ages. For these are the men that, when they have played their parts and had their 10 exits, must step out and give the moral of their scenes, and deliver unto posterity an inventory of their virtues and vices. And surely there goes a great deal of conscience to the compiling of an history. There is no reproach to the scandal of a story; it is such an authentic kind of falsehood that with authority belies our good names to all nations and posterity.

Section 4. There is another offence unto charity which no author hath ever written of, and as few take notice of; and that's the reproach not of whole professions, mysteries and conditions, but of whole nations; wherein by opprobrious 20 epithets we miscall each other, and by an uncharitable logic from a disposition in a few conclude a habit in all:

> *Le mutin Anglais, et le bravache Écossais;*
> *Le bougre Italien et le fol Français;*
> *Le poltron Romain, le larron de Gascogne,*
> *L'Espagnol superbe, et l'Allemand ivrogne.*

St Paul, that calls the Cretans liars, doth it but indirectly and upon quotation of their own poet. It is as bloody a thought in one way as Nero's was in another. For by a word we wound a thousand, and at one blow assassin the honour of 30 a nation. It is as complete a piece of madness to miscall and rave against the times, or think to recall men to reason by a

fit of passion. Democritus, that thought to laugh the times into goodness, seems to me as deeply hypochondriac as Heraclitus that bewailed them. It moves not my spleen to behold the multitude in their proper humours—that is, in their fits of folly and madness; as well understanding that wisdom is not profaned unto the world, and 'tis the privilege of a few to be virtuous. They that endeavour to abolish vice destroy also virtue; for contraries, though they destroy one another, are yet the life of one another. Thus virtue—abolish
10 vice—is an idea; again, the community of sin doth not disparage goodness, for when vice gains upon the major part, virtue—in whom it remains—becomes more excellent; and being lost in some, multiplies its goodness in others which remain untouched, and persists entire in the general inundation. I can therefore behold vice without a satire, content only with an admonition or instructive reprehension; for noble natures, and such as are capable of goodness, are railed into vice that might as easily be admonished into virtue; and we should be all so far the orators of goodness as to protect her from the
20 power of vice, and maintain the cause of injured truth. No man can justly censure or condemn another, because indeed no man truly knows another. This I perceive in myself, for I am in the dark to all the world, and my nearest friends behold me but in a cloud. Those that know me superficially think less of me than I do of myself; those of my near acquaintance think more: God, who knows me truly, knows that I am nothing, for he only beholds me and all the world who looks not on us through a derived ray or a trajection of a sensible species, but beholds the substance without the help of acci-
30 dents, and the form of things as we their operations. Further, no man can judge another, because no man knows himself; for we censure others but as they disagree from that humour which we fancy laudable in ourselves, and commend them but for that wherein they seem to quadrate and consent with

us. So that in conclusion, all is but that we all condemn, self-love. 'Tis the general complaint of these times, and perhaps of those past, that charity grows cold; which I perceive most verified in those which most do manifest the fires and flames of zeal; for it is a virtue that best agrees with coldest natures, and such as are complexioned for humility. But how shall we expect charity towards others when we are all uncharitable to ourselves? 'Charity begins at home', is the voice of the world, yet is every man his greatest enemy, and, as it were, his own executioner. *Non occides* is the commandment of God, yet scarce observed by any man; for I perceive every man is his own Atropos, and lends a hand to cut the thread of his own days. Cain was not therefore the first murderer, but Adam, who brought in death; whereof he beheld the practice only and example in his own son Abel, and saw that verified in the experience of another which faith could not persuade him in the theory of himself.

Section 5. There is, I think, no man that apprehends his own miseries less than myself, and no man that so nearly apprehends another's. I could lose an arm without a tear, and with a few groans, methinks, be quartered into pieces; yet I can weep most seriously at a play, and receive with a true passion the counterfeit griefs of those known and professed impostors. It is a barbarous part of inhumanity to add unto an afflicted party's misery, or endeavour to multiply in any man a passion whose single nature is already above his patience. This was the greatest affliction of Job, and those oblique expostulations of his friends a deeper injury than the downright blows of the devil. It is not the tears of our own eyes only but of our friends also that do exhaust the current of our sorrows, which falling into many streams runs more peaceably within its own banks, and is contented with a narrower channel. It is an act within the power of charity to translate a passion out of one breast into another, and to divide a sorrow almost out of itself; for an affliction, like a dimension, may be so divided as—if not invisible—at least to become insensible. Now, with my friend I desire not to share or participate but to engross his sorrows, that by making them mine own I may more easily discuss them; for in mine own reason and within myself, I can command that which I cannot entreat without myself and within the circle of another. I have often thought those noble pairs and examples of friendship not so truly histories of what had 10 been as fictions of what should be, but I now perceive nothing therein but possibilities, nor anything in the heroic examples of Nisus and Euryalus, Damon and Pythias, Achilles and Patroclus, which methinks upon some grounds I could not perform within the narrow compass of myself. That a man should lay down his life for his friend seems strange to vulgar affections and such as confine themselves within that worldly principle, 'Charity begins at home'. For mine own part I could never remember the relations I hold unto myself, nor the respect I owe unto mine own nature in the cause of God, 20 my country and my friends. Next to these three I do embrace myself. I confess I do not observe that order that the Schools ordain our affections, to love our parents, wives, children and then our friends; for excepting the injunctions of religion, I do not find in myself such a necessary and indissoluble sympathy to all those of my blood. I hope I do not break the fifth Commandment if I conceive I may love my friend before the nearest of my blood, even those to whom I owe the principles of life. I never yet cast a true affection on a woman, but I have loved my friend as I do virtue, and as I do my soul, my God. 30 These individual sympathies are stronger and from a more powerful hand than those specifical unions. From hence methinks I do conceive how God loves man, what happiness there is in the love of God. Omitting all other, there are three most mystical unions: two natures in one person, three persons in one nature, one soul in two bodies. For though indeed they be really divided, yet are they so united as they seem but one, and make rather a duality than two distinct souls.

Section 6. There are wonders in true affection. It is a body of enigmas, mysteries and riddles, wherein two so become one as they both become two. I love my friend before myself, and yet methinks I do not love him enough; some few months hence my multiplied affection will make me believe I have not loved him at all. When I am from him I am dead till I be with him; when I am with him I am not satisfied, but would still be nearer him. United souls are not satisfied with embraces, but desire each to be truly the other; which being impossible, their desires are infinite and must proceed without a possibility of satisfaction. Another misery there is in affection, that whom we truly love like our own we forget their looks, nor can our memory retain the idea of their faces: and it is no wonder, for they are ourselves, and our affection makes their looks our own. This noble affection falls not on vulgar and common constitutions, but on such as are marked for virtue; he that can love his friend with this noble ardour will in a competent degree affect all. Now if we can bring our affections to look beyond the body and cast an eye upon the soul, we have found out the true object not only of friendship but charity; and the greatest happiness that we can bequeath the soul is that wherein we all do place our last felicity, salvation; which though it be not in our power to bestow, it is in our charitable and pious invocations to desire, if not to procure and further. I cannot contentedly frame a prayer for my particular self without a catalogue of my friends, nor request a happiness wherein my sociable disposition doth not desire the fellowship of my neighbour. I never hear the toll of a passing bell, though in my mirth and at a tavern, without my prayers and best wishes for the departing spirit: I cannot go to cure the

body of my patient but I forget my profession, and call unto God for his soul. I cannot see one say his prayers but, instead of imitating him, I fall into a supplication for him, who, per-adventure, is no more to me than a common nature: and if God hath vouchsafed an ear to my supplications there are surely many happy that never saw me, and enjoy the blessings of mine unknown devotions. To pray for enemies—that is, for their salvation—is no harsh precept, but the practice of our daily and ordinary devotions. I cannot believe the story of
10 the Italian; our bad wishes and malevolous desires proceed no further than this life. It is the devil and uncharitable votes of hell that desire our misery in the world to come.

Section 7. To do no injury nor take none was a principle which, to my former years and impatient affections, seemed to contain enough of morality, but my more settled years and Christian constitution have fallen upon severer resolutions. I can hold there is no such thing as injury; that if there be, there is no such injury as revenge, and no such revenge as the contempt of an injury; that to hate another is to malign
20 himself; that the truest way to love another is to despise our-selves. I were unjust unto mine own conscience if I should say I am at variance with anything like myself. I find there are many pieces in this one fabric of man; and that this frame is raised upon a mass of antipathies. I am one, methinks, but as the world, wherein notwithstanding there is a swarm of dis-tinct essences, and in them another world of contrarieties. We carry private and domestic enemies within, public and more hostile adversaries without. The devil, that did but buffet St Paul, plays methinks at sharps with me. Let me
30 be nothing if within the compass of myself I do not find the battle of Lepanto; passion against reason, reason against faith, faith against the devil, and my conscience against all. There is another man within me that's angry with me, rebukes, com-

mands and dastards me. I have no conscience of marble to resist the hammer of more heavy offences, nor yet so soft and waxen as to take the impression of each single peccadillo and scape of infirmity. I am of a strange belief that it is as easy to be forgiven some sins as to commit others. For my original sin, I hold it to be washed away in my baptism; for my actual transgressions I compute and reckon with God, but from my last repentance, sacrament or general absolution; and therefore am not terrified with the sins and madness of my youth. I thank the goodness of God I have no sins that want a name. 10 I am not singular in offences; my transgressions are epidemical, and from the common breath of our corruption. For there are certain tempers of body which, matched with an humorous depravity of mind, do hatch and produce vitiosities whose newness and monstrosity of nature admits no name. This was the temper of that lecher that carnalled with a statue, and the constitution of Nero in his spintrian recreations. For the heavens are not only fruitful in new and unheard-of stars, the earth in plants and animals, but men's minds also in villainy and vices. Now the dullness of my reason and the vulgarity 20 of my disposition never prompted my invention nor solicited my affection unto any of these; yet even those common and quotidian infirmities, that so necessarily attend me and do seem to be my very nature, have so dejected me, so broken the estimation that I should have otherwise of myself, that I repute myself the most abjectest piece of mortality; that I detest mine own nature, and in my retired imaginations cannot withhold my hands from violence on myself. Divines prescribe a fit of sorrow to repentance. There goes indignation, anger, contempt and hatred into mine, passions of a contrary nature 30 which neither seem to suit with this action nor my proper constitution. It is no breach of charity to ourselves to be at variance with our vices, nor to abhor that part of us which is an enemy to the ground of charity, our God; wherein we do but imitate our great selves, the world, whose divided anti-pathies and contrary faces do yet carry a charitable regard unto the whole; by their particular discords preserving the common harmony, and keeping in fetters those powers whose rebellions, once masters, might be the ruin of all.

Section 8. I thank God, amongst those millions of vices I do inherit and hold from Adam I have escaped one, and that a mortal enemy to charity, the first and father sin not only of man but of the devil, pride; a vice whose name is compre-
10 hended in a monosyllable, but in its nature not circumscribed with a world. I have escaped it in a condition that can hardly avoid it. Those petty acquisitions and reputed perfections that advance and elevate the conceits of other men add no feathers unto mine. I have seen a grammarian tower and plume himself over a single line in Horace, and show more pride in the construction of one ode than the author in the composure of the whole book. For my own part, besides the jargon and patois of several provinces, I understand no less than six languages; yet I protest I have no higher conceit of myself
20 than had our fathers before the confusion of Babel, when there was in the world but one language and none to boast himself either linguist or critic. I have not only seen several countries, beheld the nature of their climes, the chorography of their provinces, topography of their cities, but understand their several laws, customs and policies; yet cannot all this persuade the dullness of my spirit unto such an opinion of myself as I behold in nimbler and conceited heads, that never looked a degree beyond their nests. I know the names and somewhat more of all the constellations in my horizon, yet
30 I have seen a prating mariner that could only name the Pointers and the North Star out-talk me, and conceit himself a whole sphere above me. I know most of the plants of my country and of those about me; yet methinks I do not know

so many as when I did but know an hundred, and had scarcely ever simpled further than Cheapside. For indeed, heads of capacity, and such as are not full with a handful or easy measure of knowledge, think they know nothing till they know all; which being impossible, they fall upon the opinion of Socrates, and only know they know not anything. I cannot think that Homer pined away upon the riddle of the fisherman, or that Aristotle, who understood the uncertainty of knowledge and confessed so often the reason of man too weak for the works of nature, did ever drown himself upon the flux 10 and reflux of Euripus. We do but learn today what our better advanced judgements will unteach us tomorrow; and Aristotle doth but instruct us as Plato did him—that is, to confute himself. I have run through all sects, yet find no rest in any. Though our first studies and junior endeavours may style us Peripatetics, Stoics or Academics, yet I perceive the wisest heads prove at last almost all Sceptics, and stand like Janus in the field of knowledge. I have therefore one common and authentic philosophy I learned in the Schools whereby I discourse and satisfy the reason of other men; another more 20 reserved and drawn from experience whereby I content mine own. Solomon, that complained of ignorance in the height of knowledge, hath not only humbled my conceits but discouraged my endeavours. There is yet another conceit that hath sometimes made me shut my books, which tells me it is a vanity to waste our days in the blind pursuit of knowledge. It is but attending a little longer, and we shall enjoy that by instinct and infusion which we endeavour at here by labour and inquisition. It is better to sit down in a modest ignorance and rest contented with the natural blessing of our own 30 reasons than buy the uncertain knowledge of this life with sweat and vexation, which death gives every fool gratis, and is an accessory of our glorification.

Section 9. I was never yet once, and commend their resolutions who never marry twice; not that I disallow of second marriage as neither, in all cases, of polygamy; which considering some times and the unequal number of both sexes, may be also necessary. The whole woman was made for man, but the twelfth part of man for woman. Man is the whole world and the breath of God; woman the rib and crooked piece of man. I could be content that we might procreate like trees, without conjunction, or that there were any way to per-
10 petuate the world without this trivial and vulgar way of coition. It is the foolishest act a wise man commits in all his life; nor is there anything that will more deject his cooled imagination when he shall consider what an odd and unworthy piece of folly he hath committed. I speak not in prejudice, nor am I averse from that sweet sex, but naturally amorous of all that is beautiful. I can look a whole day with delight upon a handsome picture, though it be but of an horse. It is my temper—and I like it the better—to affect all harmony; and sure there is a music, even in the beauty and
20 the silent note which Cupid strikes, far sweeter than the sound of an instrument. For there is a music wherever there is a harmony, order or proportion; and thus far we may maintain the music of the spheres. For those well-ordered motions and regular paces, though they give no sound unto the ear, yet to the understanding they strike a note most full of harmony. Whosoever is harmonically composed delights in harmony; which makes me much mistrust the symmetry of those heads which declaim against all church music. For myself, not only from my obedience but my particular genius I do embrace it:
30 for even that vulgar and tavern music which makes one man merry, another mad, strikes me into a deep fit of devotion and a profound contemplation of the first composer. There is something in it of divinity more than the ear discovers. It is a hieroglyphical and shadowed lesson of the whole world and

the creatures of God; such a melody to the ear as the whole
world well understood would afford the understanding. In
brief, it is a sensible fit of that harmony which intellectually
sounds in the ears of God. It unties the ligaments of my frame,
takes me to pieces, dilates me out of myself, and by degrees,
methinks, resolves me into heaven. I will not say with Plato
the soul is an harmony, but harmonical, and hath its nearest
sympathy unto music. Thus some whose temper of body
agrees and humours the constitution of their souls are born
poets, though indeed all are naturally inclined unto rhythm. 10
This made Tacitus, in the very first line of his story, fall upon
a verse; and Cicero, the worst of poets, but declaiming for a
poet, fall in the very first sentence upon a perfect hexameter.
I feel not in me those sordid and unchristian desires of my profession; I do not secretly implore and wish for plagues, rejoice
at famines, revolve ephemerides and almanacs in expectation
of malignant aspects, fatal conjunctions and eclipses. I rejoice
not at unwholesome springs nor unseasonable winters. My
prayers go with the husbandman's; I desire everything in its
proper season, that neither men nor the times be out of 20
temper. Let me be sick myself if sometimes the malady of my
patient be not a disease unto me; I desire rather to cure his
infirmities than my own necessities. Where I do no good
methinks it is scarce honest gain, though I confess 'tis but the
worthy salary of our well-intended endeavours. I am not only
ashamed but heartily sorry that besides death there are diseases
incurable; yet not for my own sake, or that they be beyond
my art, but for the general cause and sake of humanity whose
common cause I apprehend as mine own. And to speak more
generally, those three noble professions which all civil com- 30
monwealths do honour are raised upon the fall of Adam, and
are not any way exempt from their infirmities. There are not
only diseases incurable in physic but cases indissoluble in laws,
vices incorrigible in divinity. If general councils may err, I do

not see why particular courts should be infallible. Their perfectest rules are raised upon the erroneous reason of man, and the laws of one do but condemn the rules of another; as Aristotle ofttimes the opinions of his predecessors, because, though agreeable to reason, yet were not consonant to his own rules and the logic of his proper principles. Again—to speak nothing of the sin against the Holy Ghost, whose cure not only but whose nature is unknown—I can cure the gout or stone in some, sooner than divinity pride or avarice in others.
10 Further, I can cure vices by physic when they remain incurable by divinity, and they shall obey my pills when they contemn their precepts. I boast nothing, but plainly say we all labour against our own cure, for death is the cure of all diseases. There is no catholicon or universal remedy I know but this; which though nauseous to queasier stomachs, yet to prepared appetites is nectar, and a pleasant potion of immortality.

Section 10. For my conversation, it is like the sun's, with all men, and with a friendly aspect to good and bad. Methinks
20 there is no man bad, and the worst, best—that is, while they are kept within the circle of those qualities wherein they are good. There is no man's mind of such discordant and jarring a temper to which a tunable disposition may not strike a harmony. *Magnae virtutes nec minora vitia*; it is the posy of the best natures and may be inverted on the worst. There are in the most depraved and venomous dispositions certain pieces which remain untouched; which by an antiperistasis become more excellent, or by the excellency of their antipathies are able to preserve themselves from the contagion of their
30 enemy vices, and persist entire beyond the general corruption. For it is also thus in nature. The greatest balsams do lie enveloped in the bodies of most powerful corrosives. I say, moreover, and I ground upon experience, that poisons contain

within themselves their own antidote, and that which preserves them from the venom of themselves, without which
they were not only deleterious to others but to themselves also.
But it is the corruption that I fear within me, not the contagion
of commerce without me. 'Tis that unruly regiment within
me that will destroy me; 'tis I that do infect myself. The man
without a navel yet lives in me; I feel that original canker
corrode and devour me; and therefore *Defenda me Dios de me*,
Lord deliver me from myself, is a part of my litany, and the
first voice of my retired imaginations. There is no man alone, 10
because every man is a microcosm and carries the whole world
about him. *Nunquam minus solus quam cum solus*, though it be
the apothegm of a wise man is yet true in the mouth of a fool.
For indeed, though in a wilderness a man is never alone, not
only because he is with himself and his own thoughts, but
because he is with the devil, who ever consorts with our solitude, and is that unruly rebel that musters up those disordered
motions which accompany our sequestered imaginations. And
to speak more narrowly, there is no such thing as solitude, nor
anything that can be said to be alone and by itself but God, who 20
is his own circle, and can subsist by himself. All others—
besides their dissimilar and heterogeneous parts, which in a
manner multiply their natures—cannot subsist without the concourse of God and the society of that hand which doth uphold
their natures. In brief, there can be nothing truly alone and by
itself which is not truly one, and such is only God. All others
do transcend an unity, and so by consequence are many.

Section 11. Now for my life, it is a miracle of thirty years,
which to relate were not an history but a piece of poetry,
and would sound to common ears like a fable. For the world, 30
I count it not an inn but an hospital, and a place not to live
but to die in. That world which I regard is myself; it is the
microcosm of mine own frame that I cast mine eye on; for

the other, I use it but like my globe, and turn it round sometimes for my recreation. Men that look upon my outside, perusing only my condition and fortunes, do err in my altitude; for I am above Atlas' shoulders, and though I seem on earth to stand, on tiptoe in heaven. The earth is a point, not only in respect of the heavens above us but of that heavenly and celestial part within us. That mass of flesh that circumscribes me limits not my mind: that surface that tells the heavens it hath an end cannot persuade me I have any.
10 I take my circle to be above three hundred and sixty. Though the number of the arc do measure my body, it comprehendeth not my mind. Whilst I study to find how I am a microcosm or little world, I find myself something more than the great. There is surely a piece of divinity in us; something that was before the elements, and owes no homage unto the sun. Nature tells me I am the image of God as well as Scripture. He that understands not thus much hath not his introduction or first lesson, and is yet to begin the alphabet of man. Let me not injure the felicity of others if I say I am as happy
20 as any. I have that in me that can convert poverty into riches, transform adversity into prosperity: I am more invulnerable than Achilles; fortune hath not one place to hit me. *Ruat coelum, fiat voluntas tua*, salveth all; so that whatsoever happens, it is but what our daily prayers desire. In brief, I am content, and what should providence add more? Surely this is it we call happiness, and this do I enjoy. With this I am happy in a dream, and as content to enjoy a happiness in a fancy, as others in a more apparent truth and reality. There is surely a nearer apprehension of anything that delights us in our dreams than
30 in our awaked senses. With this I can be a king without a crown, rich without a royalty, in heaven though on earth, enjoy my friend and embrace him at a distance; without which I cannot behold him. Without this I were unhappy; for my awaked judgement discontents me, ever whispering

unto me that I am from my friend; but my friendly dreams in the night requite me, and make me think I am within his arms. I thank God for my happy dreams as I do for my good rest; for there is a satisfaction in them unto reasonable desires, and such as can be content with a fit of happiness; and surely it is not a melancholy conceit to think we are all asleep in this world, and that the conceits of this life are as mere dreams to those of the next, as the phantasms of the night to the conceits of the day. There is an equal delusion in both, and the one doth but seem to be the emblem and picture of the other. We are somewhat more than ourselves in our sleeps, and the slumber of the body seems to be but the waking of the soul. It is the ligation of sense, but the liberty of reason; our waking conceptions do not match the fancies of our sleeps. At my nativity my ascendant was the watery sign of Scorpius: I was born in the planetary hour of Saturn, and I think I have a piece of that leaden planet in me. I am no way facetious, nor disposed for the mirth and galliardise of company, yet in one dream I can compose a whole comedy, behold the action, apprehend the jests and laugh myself awake at the conceits thereof. Were my memory as faithful as my reason is then fruitful I would never study but in my dreams, and this time also would I choose for my devotions; but our grosser memories have then so little hold of our abstracted understandings that they forget the story, and can only relate to our awaked souls a confused and broken tale of what hath passed. Aristotle, who hath written a singular tract of sleep, hath not methinks throughly defined it, nor yet Galen, though he seem to have corrected it; for those noctambuloes and nightwalkers, though in their sleep, do yet enjoy the action of their senses. We must therefore say that there is something in us that is not in the jurisdiction of Morpheus, and that these abstracted and ecstatic souls do walk about in their own corpse as spirits with the bodies they assume, wherein they seem to hear,

see and feel, though indeed the organs are destitute of sense, and
their natures of those faculties that should inform them. Thus it
is observed that men sometimes, upon the hour of their depar-
ture, do speak and reason above themselves; for then the soul,
beginning to be freed from the ligaments of the body, begins to
reason like herself and to discourse in a strain above mortality.

Section 12. We term sleep a death, and yet it is waking that
kills us and destroys those spirits which are the house of life.
'Tis indeed a part of life that best expresseth death, for every
man truly lives so long as he acts his nature, or some way
makes good the faculties of himself. Themistocles, therefore,
that slew his soldier in his sleep, was a merciful executioner.
'Tis a kind of punishment the mildness of no laws hath invented: I wonder the fancy of Lucan and Seneca did not discover it. It is that death by which we may be literally said to die
daily, a death which Adam died before his mortality; a death
whereby we live a middle and moderating point between life
and death; in fine, so like death I dare not trust it without my
prayers and an half adieu unto the world. And truly 'tis a
fit time for our devotion; and therefore I cannot lay down my
head without an orison, and take my farewell in a colloquy
with God.

> The night is come like to the day,
> Depart not thou great God away.
> Let not my sins, black as the night,
> Eclipse the lustre of thy light.
> Keep still in my horizon, for to me
> The sun makes not the day, but thee.
> Thou whose nature cannot sleep,
> On my temples sentry keep;
> Guard me 'gainst those watchful foes,
> Whose eyes are open while mine close.
> Let no dreams my head infest,
> But such as Jacob's temples blessed.

While I do rest, my soul advance,
Make my sleep a holy trance:
That I may—my rest being wrought—
Awake into some holy thought;
And with as active vigour run
My course as doth the nimble sun.
Sleep is a death; O make me try
By sleeping, what it is to die,
And down as gently lay my head
On my grave, as now my bed.　　　　　　10
Howe'er I rest, great God, let me
Awake again at last with thee.
And thus assured, behold I lie
Securely, or to wake or die.
These are my drowsy days, in vain
I now do wake to sleep again.
O come that hour, when I shall never
Sleep again, but wake for ever!

This is the dormitive I take to bedward; I need no other
laudanum than this to make me sleep. After which I close 20
mine eyes in security, content to take my leave of the sun and
sleep unto the Resurrection.

Section 13. The method I should use in distributive justice
I often observe in commutative, and keep a geometrical proportion in both; whereby, becoming equable to others, I
become unjust to myself, and supererogate in that common
principle, Do unto others as thou wouldest be done unto
thyself. I was not born unto riches, nor is it, I think, my star
to be wealthy; or, if it were, the freedom of my mind and
frankness of my disposition were able to contradict and cross 30
my fates. For to me avarice seems not so much a vice as a
deplorable piece of madness. To conceive ourselves urinals
or be persuaded that we are dead is not so ridiculous, nor so
many degrees beyond the power of hellebore as this. The

opinions of theory and positions of men are not so void of reason as their practised conclusion. Some have held that snow is black, that the earth moves, that the soul is fire, air, water; but all this is philosophy, and there is no delirium—if we do but speculate—to the folly and indisputable dotage of avarice. To that subterraneous idol and God of the earth I do confess I am an atheist. I cannot persuade myself to honour that which the world adores. Whatsoever virtue its prepared substance may have within my body, it hath no influence nor
10 operation without. I would not entertain a base design, or an action that should call me villain, for the Indies; and for this only do I love and honour my own soul, and have methinks two arms too few to embrace myself. Aristotle is too severe that will not allow us to be truly liberal without wealth and the bountiful hand of fortune. If this be true, I must confess I am only charitable in my liberal intentions and bountiful good wishes. But if the example of the mite be not only an act of wonder but an example of the noblest charity, surely poor men may also build hospitals, and the rich alone have not
20 erected cathedrals. I have a private method which others observe not. I take the opportunity of myself to do good: I borrow occasion of charity from mine own necessities, and supply the wants of others when I am most in need myself. When I am reduced to the last tester I love to divide it with the poor; for it is an honest stratagem to take advantage of ourselves, and so to husband the acts of virtue that where they are defective in one circumstance they may repay their want and multiply their goodness in another. I have not Peru in my desires, but a competence and ability to perform those good
30 works to which he hath inclined my nature. He is rich who hath enough to be charitable; and it is hard to be so poor where a noble mind may not find a way to this piece of goodness. *He that giveth to the poor lendeth to the Lord*: there is more rhetoric in that one sentence than in a library of sermons.

And indeed, if these sentences were understood by the reader
with the same emphasis as they are delivered by the author, we
needed not these volumes of instructions, but might be honest
by an epitome. Upon this motive only I cannot behold a beggar
without relieving his necessities with my purse or his soul with
my prayers. These scenical and accidental differences between
us cannot make me forget that common and untouched part of
us both. There is, under these centoes and miserable outsides,
these mutilate and semi-bodies, a soul of the same alloy with
our own, whose genealogy is God as well as ours, and in as 10
fair a way unto salvation as ourselves. Statists that labour to
contrive a commonwealth without poverty take away the
object of charity, not understanding only the commonwealth
of a Christian but forgetting the prophecy of Christ.

Section 14. Now there is another part of charity which is the
basis and pillar of this, and that is the love of God, for whom
we love our neighbour: for this I think is charity, to love God
for himself and our neighbour for God. All that is truly
amiable is God, or as it were a divided piece of him that retains
a reflex or shadow of himself. Nor is it strange that we should 20
place affection on that which is invisible. All that we truly
love is thus. What we adore under affection of our senses
deserves not the honour of so pure a title. Thus we adore
virtue, though to the eyes of sense she be invisible. Thus that
part of our noble friends that we love is not the part we
embrace, but that insensible part that our arms cannot embrace. God, being all goodness, can love nothing but himself.
He loves us but for that part which is as it were himself, and
the traduction of his holy spirit. Let us call to assize the loves
of our parents, the affection of our wives and children, and 30
they are all dumb-shows and dreams, without reality, truth
or constancy. For first there is a strong bond of affection
between us and our parents, yet how easily dissolved! We

betake ourselves to a woman and forget our mother in a wife, the womb that bare us in that that shall bear our image. This woman blessing us with children, our affection leaves the level it held before, and sinks from our bed unto our issue and pictures of posterity, where affection holds no steady mansion. They growing up in years either desire our ends, or, applying themselves to a woman, take a lawful way to love another better than ourselves. Thus I perceive a man may be buried alive, and behold his grave in his own issue.

10 *Section 15.* I conclude therefore and say there is no happiness under—or, as Copernicus will have it, above—the sun, nor any cramb in that repeated verity and burden of all the wisdom of Solomon, *All is vanity and vexation of spirit.* There is no felicity in what the world adores. Aristotle, whilst he labours to refute the Ideas of Plato, falls upon one himself; for his *summum bonum* is a chimera, and there is no such thing as his felicity. That wherein God himself is happy, the holy angels are happy, in whose defect the devils are unhappy, that dare I call happiness. Whatsoever conduceth unto this may
20 with an easy metaphor deserve that name; whatsoever else the world terms happiness is to me a story out of Pliny, a tale of Boccace or Malizspini, an apparition or neat delusion wherein there is no more of happiness than the name. Bless me in this life with but the peace of my conscience, command of my affections, the love of thyself and my dearest friends, and I shall be happy enough to pity Caesar. These are, O Lord, the humble desires of my most reasonable ambition, and all I dare call happiness on earth: wherein I set no rule or limit to thy hand or providence. Dispose of me according to the
30 wisdom of thy pleasure. Thy will be done, though in my own undoing.

FINIS.

Triuni Deo sit gloria in aeternum. Amen.

COMMENTARY AND NOTES

The following abbreviated references are used:

Carter   *Urn Burial,* and *The Garden of Cyrus,* by Sir Thomas Browne, edited by John Carter; Cambridge University Press, 1958.

Cross   *Dictionary of the Christian Church,* edited by F. L. Cross; Oxford University Press, 1957.

Denonain   *Religio Medici,* edited by J.-J. Denonain; Cambridge University Press, 1955.

Digby   *Observations upon Religio Medici,* by Sir Kenelm Digby; London, 1643.

Greenhill   *Religio Medici,* edited by W. A. Greenhill; London, 1881.

Hakewill   *Apology for the power and providence of God,* by George Hakewill; Oxford, 1627.

Healey   *The City of God,* by St Augustine; translated by John Healey, 1610.

Keynes   *The Letters of Sir Thomas Browne,* edited by Geoffrey Keynes; London, 1941.

McKerrow   *The Works of Thomas Nashe,* edited by R. B. McKerrow; Oxford, 1958.

Montaigne   *Essais,* edited by Maurice Rat; Paris, 1958.

Pseud. Epid. Pseudodoxia Epidemica, by Sir Thomas Browne; London, 1646.

Sanna   *Religio Medici,* edited by Vittoria Sanna, Cagliari, 1958.

PAGE I

7 *perversion of that excellent invention* printing; by the publication of anti-royalist tracts before the outbreak of civil war in 1642.

9 *complaints* against the pirating of *Religio Medici* by Andrew Crooke in 1642.

97

11 *incapable of affronts* too negligible to be insulted.

20-1 *full and intended copy* According to Denonain, Browne corrected some 650 mistakes in the pirated text, but overlooked more than 350 others.

that piece here meaning work, and a favourite word with Browne; see Section 15, p. 18, in which it recurs six times.

23 *about seven years past* about 1635, when Browne was living at Upper Shibden Hall, near Halifax.

PAGE 2

10 *plausible unto my past apprehension* agreeable to my former outlook.

16 *as I have declared* see p. 70, ll. 3-6.

PAGE 3

3 *the general scandal of my profession* that most doctors were atheists. This slur dates at least from the fourteenth century; see Chaucer's comment on his Doctor of Physic, 'His studie was but litel on the Bible' (*The General Prologue*, 438).

3-4 *the natural course of my studies* Anatomy was still regarded as an impious prying unto divine handiwork.

14 *the principles of grace* 'The supernatural assistance of God bestowed upon a rational being with a view to his sanctification' (Cross).

25-6 *reformed new-cast religion* the Protestant faith. Browne dislikes the connotation of protesting which the term carries.

28 *the Fathers* the early Christian writers; usually those of the first five centuries.

PAGE 4

7 *the person* Martin Luther, 1483-1546, the son of a miner from Saxony. He was moved to contest the authority of the Pope in 1517, when indulgences were offered to those who contributed to the repair of St Peter's in Rome: the 'accidental occasion' which Browne refers to. Within a few weeks his protest was known and supported throughout Germany.

9-10 *same objection...cast at Christ* Mark vi. 2-3.

11 *shaken hands with* taken leave of, dissociated myself from.

11–12 *those desperate resolutions* the Roman Catholics.

16 *in diameter* in opposition; repeated on p. 62, l. 17.

PAGE 5

15 *Ave Marie bell* 'A church bell that tolls every day at 6 and 12 of the clock, at the hearing whereof everyone in what place soever either of house or street betakes him to his prayer, which is commonly directed to the Virgin' (marginal gloss, 1643).

PAGE 6

16 *subscribe unto her articles* But in Section 60 Browne contests justification by faith, without appearing to notice that he is denying the eleventh of the Thirty-nine Articles.

23 *Calvin* John Calvin, 1509–64, French reformer and theologian. From 1536 he helped to organise the Reformation at Geneva.

24 *Council of Trent, Synod of Dort* held respectively at Trento, Northern Italy, from 1545 to 1563, and at Dordrecht in Holland during 1618–19. The first was a universal council of Roman Catholic authorities which determined or re-established essential points of religious doctrine; the second an assembly of the Dutch Reformed Church which ended in a victory for the Calvinists.

33 *the state of Venice would have attempted* referring to a dispute between the Venetian Republic and Pope Paul V in 1606, which was settled a year later.

PAGE 7

16–18 *or be angry...dissent myself* 'I cannot think but in this expression the author had respect to that of that excellent French writer M. Montaigne, in which I often trace him' (Digby). The passage runs: 'Ce que je tiens aujourd'huy et ce que je croy, je le tiens et le croy de toute ma croyance...mais ne m'est-il pas advenu, non une fois, mais cent, mais mille, et tous les jours, d'avoir ambrassé quelqu'autre chose à tout ces mesmes instrumens, en cette mesme condition, que depuis j'aye jugée fauce?' (*Essais*, II, 267–8).

20 *especially upon a disadvantage* when the opponent has a stronger argument.

27–8   *take up the gauntlet* accept a challenge.

PAGE 8

 4   *Oedipus* who delivered Thebes by solving the riddle of the Sphinx.

 8   *no man more paradoxical* enjoying paradoxes and enigmas.

11–12   *not reserving…my own brain* without any reservations suggested by my private ruminations.

15   *greener* those of a young man.

20   *the river Arethusa* 'that loseth itself in Greece and riseth again in Sicily' (marginal gloss, 1643).

23–4   *the like aspects from heaven* referring to the Platonic year, explained below.

25   *metempsychosis* a belief advocated by Plato and Pythagoras, that the human soul migrates into another body at death, continuing its existence on earth until it attains to complete spiritual purity.

29   *Plato's year* 'A revolution of certain thousand years when all things should return unto their former estate, and he be teaching again in his school as when he delivered this opinion' (marginal gloss). At the completion of this period all the heavenly bodies would have returned to their initial positions, and the astrological influences, or aspects, would be the same as at first.

30   *Diogenes, Timon* the first a Cynic philosopher, the other a misanthropist, both of whom shunned the company of men.

PAGE 9

 1   *the first of mine* my first heretical belief.

12   *I could with patience be nothing* see Digby's remark on 'above Atlas' shoulders', p. 90, l. 4.

13   *Origen* a Christian theologian and writer of the third century, who held that all creatures, even the devil, would eventually be saved.

PAGE 10

4–7   *Those have not…be of a sect also* they must have disordered minds and feelings who cannot happily dissent from orthodox belief unless it is heretical to do so, or express their private opinions without the support of a party.

13–14 *the prophecy of Christ* Matthew xxiv. 24. See also p. 56, ll. 2–3.

17 *Arians* followers of Arius, a fourth-century presbyter who denied that Christ was co-substantial with God.

25 *men of singular parts* gifted or otherwise outstanding individuals.

31 *the Schools* Scholastic philosophy; devised by the medieval Schoolmen upon a basis of Aristotelian thought and Patristic theology.

PAGE II

3 *pia mater* the innermost of three membranes enclosing the brain, often used, as here, to refer to the brain itself.

8 *O altitudo* part of the Vulgate of Romans xi. 33: 'O the depth of the riches both of the wisdom and knowledge of God! how unsearchable are his judgements, and his ways past finding out!'

12 *Tertullian* a third-century Father of the African Church, and the first Christian theologian to write in Latin.

   Certum est quia impossibile est 'Yea, but if, because it is wonderful it be therefore not believed, it ought on that account rather to be believed' (Greenhill).

16 *the miracle* of the Red Sea crossing; Exodus xiv. 21–31.

19 *I would not have been* I do not wish that I had been.

22 *that greater blessing* John xx. 29.

27–8 *we owe this faith unto history* we must believe Holy Writ, as a record of historical events.

28 *they only had* only those who lived before Christ's coming.

30 *mystical types* symbols foreshadowing the Incarnation.

32 *an edge* a cutting edge.

PAGE 12

1 *the Apostle* St Paul; Ephesians vi. 16.

3–4 *to know we know nothing* compare p. 85, ll. 2–6, and both with p. 65, l. 33–p. 66, l. 1.

6 *platonic description* a general definition involving mystical concepts.

7 *that allegorical description* 'sphaera cujus centrum ubique, cir-

cumferentia nullibi' (marginal gloss)—a circle whose centre is everywhere, and circumference nowhere.

Hermes Hermes Trismegistus; the name given by students of mysticism and alchemy to the Egyptian god regarded as the author of all occult doctrines.

10–11 *that anima est...perspicui* that the soul is the angel of the man, is the body of God, rather than that which naturally makes the body to move; that light is the shadow of God rather than that it is visible movement: meaning that Browne prefers mystical or paradoxical ideas to scientific definitions.

14 *how unable it is to display* how incapable reason is of explaining natural phenomena to the understanding.

16–17 *my haggard...unto the lure* a metaphor from hawking; calling back an undisciplined bird—reason—to compliance with faith.

19 *the same chapter* Genesis ii. 5.

22–3 *if we shall literally understand it* if the tempter of Eve was a serpent, he moved on his belly before God cursed him.

25 *which God ordained* Deuteronomy xxii. 13–17.

30–2 *this I think is no vulgar part of faith...contrary to reason* 'Only in the later and decadent Scholasticism was it maintained that one and the same proposition could be proved untrue by reason and accepted as true by faith' (Cross).

PAGE 13

1 *retired imagination* private contemplation; repeated on p. 83, l. 27 and p. 89, l. 10.

1–2 *neque enim...mihi* not even when I am relaxing do I forget myself. Adapted, or misquoted, from Horace, *Satires*, I, iv, 133 f., which has *excepit* for *accipit*.

8–9 *five days older than ourselves* since man was created on the sixth day.

13 *St Paul's sanctuary* a second allusion to Romans xi. 33; Browne taking refuge in the same astonishment when he contemplates the incomprehensibility of God.

20 *that terrible term predestination* properly, 'the divine decree according to which certain persons are infallibly guided to eternal salvation' (Cross). Browne appears to accept the term in its

Calvinist interpretation, which rejects the universal saving will of God and asserts that salvation is denied to the damned from all eternity without any fault on their part.

27 *St Peter spake* II Peter iii. 8.

31 *what to us is to come, to his eternity is present* 'He sees them not by change in thought, but immutably, be they past or not past, to come or not to come: all these he has eternally present...all things he knows are in his unbodily presence' (Healey, XI, 21).

PAGE 14

3 *deny a priority* suppose that neither Father nor Son takes precedence.

4 *Aristotle* Greek philosopher, 384–322 B.C. With Plato, the outstanding thinker of the ancient world; whose works on metaphysics, natural science, politics, ethics and religion provided the basis of medieval Scholasticism.

5 *make good* support a belief in.

9 *three distinct souls* vegetative, animal and rational: a belief of the Peripatetic philosophers, who supposed that man combined with his own characteristic faculties and intelligence those of the plants and the creatures.

different faculties animal, vital and natural: meaning those of the brain and the nervous system, of the heart and lungs, and of nutriment and assimilation.

16 *Pythagoras* Greek philosopher and mathematician of the sixth century B.C. who 'did suppose numbers to be the principles and original of things' (Bacon, *Advancement of Learning*, II, viii, 1). He was the first to propose the theory of metempsychosis, referred to in Section 37, p. 46, l. 4.

the secret magic of numbers such as Browne explores in his study of the quincunx in *The Garden of Cyrus*. 'To enlarge this contemplation unto all the mysteries and secrets accommodable unto this number (five), were inexcusable Pythagorism' (Carter p. 108).

17 *beware of philosophy* Colossians ii. 8.

18 *this mass of nature* the created world.

25 *that this visible world...invisible* Hermetic philosophy attracted

much attention from the devotional poets of Browne's age. For the influence of this particular idea see, for example, Vaughan's *The Waterfall.*

26 *things are not truly* physical objects have no real existence.

PAGE 15

10 *as he did at Delphos* γνῶθι σεαυτόν—know thyself—was inscribed in gold on the wall of the temple of Apollo at Delphi. Browne associates all pagan deities with the devil.

   we had better known ourselves we should have known ourselves better.

14 *Moses' eye* Exodus xxxiii. 23.

22–3 *his actions...with deliberation* God does not need to deliberate over his course of action, but knows without reasoning.

25 *purest Ideas of goodness* associating a Platonic concept of ultimate reality with orthodox religious belief: see p. 74, ll. 32–3 and p. 102, l. 15.

30–1 *the obvious effects of nature* in visible natural phenomena.

32 *sanctum sanctorum* holy of holies.

PAGE 16

7–8 *deliberate research of his creatures* thoughtful enquiry into the works of God.

16 *meteors* fiery exhalations.

18–19 *Teach my endeavours...I may proceed* Browne wishes to study the natural world in order to understand God: the purpose of the old science, which was not concerned to provide man with useful or practical knowledge.

34–5 *Lord, lord...Father* Matthew vii. 21.

35–p. 17, l. 1 *our wills must be our performances* our wish to do God's will must be regarded as the achievement of that purpose.

PAGE 17

5 *first cause* in Aristotelian and Scholastic philosophy, the First Mover or Creator of the universe. The four secondary causes derived from this are: (i) the efficient cause, the force or agency by which a thing is produced; (ii) the formal cause, the form or

essential qualities of the thing produced; (iii) the material cause, or matter from which it is produced, and (iv) the final cause, or end for which it is produced.

7–8 *the first matter* matter without form, from which the universe was created; compare p. 43, l. 26.

9 *some positive end* some particular function or purpose within the created universe.

22 *De usu partium* Galen, who died in 199, was one of the most distinguished physicians of ancient Rome. His description of the working of the human body went unchallenged until modern times. This treatise on the use of the parts of the body includes praise and veneration for its maker.

Suarez Francisco de Suarez, 1548–1617, a Spanish Jesuit theologian. His *Disputationes Metaphysicae*, published in 1597, combined the teaching of Aristotle with that of Aquinas and exerted considerable influence upon contemporary Protestant thought.

23 *the enquiry of this cause* to understand God's purpose as well as the nature of the occurrence. 'The handling of final causes...hath intercepted the severe and diligent enquiry of all real and physical causes, and given men the occasion to stay upon these satisfactory and specious causes, to the great arrest and prejudice of further discovery' (Bacon, *Advancement of Learning*, II, vi, 7).

26 *Natura nihil agit frustra* nothing which nature does is useless.

27 *no grotesques in nature* 'so M. Montaigne: "Il n'y a rien d' inutile en nature; non pas l'inutilité mesmes"' (Digby; Montaigne, III, 2). Sanna points to the probable source of the idea in Galen, *De usu partium*, XIII, 4, 78.

30 *seeds and principles* Cf. Macbeth's 'treasure of nature's germens', IV, i, 58–9. The New Arden editor refers to a passage in the *De Trinitate* of St Augustine: 'some hidden seeds of all things that are born corporeally and visibly are concealed in the corporea elements of this world.'

PAGE 18

1 *Solomon chose...admiration* Proverbs vi. 6–8.

3 *What wise hand...?* 'Pourquoy espessit l'araignée sa toile en un endroit et relasche en un autre? se sert à cette heure de cette sorte

de neud, tantost de celle-là, si elle n'a et deliberation, et pense-
ment, et conclusion?' (Montaigne, II, 136).

10 *Regio-Montanus' fly* Johannes Müller of Königsberg, a fifteenth-
century craftsman, was reputed to have constructed an iron fly
and a wooden eagle, both capable of independent flight. Hakewill
refers to these machines in his *Apology* as proof of the un-
diminished ingenuity of modern man.

11 *two souls in those little bodies* vegetative and animal, in the insects.

17 *without further travel* probably travail, labour, is meant; but both
senses may be present.

20 *which he that studies wisely* 'For the wit and mind of man, if it...
work upon itself, as the spider worketh his web, then it is endless,
and brings forth indeed cobwebs of learning, admirable for the
fineness of thread and work, but of no substance or profit'
(Bacon, *Advancement of Learning*, I, iv, 5). See also p. 89, ll. 32–3.

21 *in a compendium* the first of Browne's many references to man as
a microcosm, which he repeats on pp. 42, 48, 61, 89 and 90. The
idea had come under attack from Bacon: 'The ancient opinion
that man was *microcosmos*, an abstract or model of the world, hath
been fantastically strained by Paracelsus and the alchemists, as if
there were to be found in man's body certain correspondences
and parallels, which should have respect to all varieties of things,
as stars, planets, minerals, which are extant in the great world'
(*The Advancement of Learning*, II, x, 2). It is characteristic of
Browne's mental disposition that he clings so persistently to the
older and more picturesque tradition of thought.

23 *two books* Holy Writ and the world of nature, which is God's
handiwork.

29 *its supernatural station* 'So the sun stood still in the midst of
heaven, and hasted not to go down about a whole day' (Joshua
x. 13).

PAGE 19

4 *the principle of motion and rest* according to Aristotelian physics,
'motus rerum est rapidus extra locum, placidus in loco'. An
object not occupying its natural place would tend to move
rapidly towards it; once there, it would remain inert. The

example of a stone dropped from a tower illustrates this natural law.

7–8   *To make a revolution...sun* an example of a final cause.

14   *sweetened the water with a wood* Exodus xv. 25.

16   *for God is like...* Browne's dogmatic assertions in this passage conflict with his admission of man's inability to understand God on p. 15, ll. 14 ff., which itself precedes a further definition of divine nature.

33–4   *outward shapes and figures...inward forms* a familiar point of Neoplatonist doctrine. 'Judge you how plainly in the face of a lion, a horse, and an eagle, a man shall discern anger, fierceness, and stoutness; in lambs and doves, simpleness and very innocency; the crafty subtlety in foxes and wolves' (Castiglione, *The Courtier*, tr. Hoby).

PAGE 20

4–5   *there is therefore no deformity...a kind of beauty* not a very happy example of logical thinking. The phrase 'wherein notwithstanding', repeated four lines later, shows Browne struggling to clear himself from a tangle of paradoxical notions and to re-establish the thread of his former argument.

15–16   *all things are artificial* that is, made by process of art or artifice. This commonplace was much repeated during the seventeenth century, when the new scientists were looking for authority for their questioning of natural phenomena.

17   *This is the...providence* Browne now reverts to the topic which —in order to renew his discussion of the theme 'there are no grotesques in nature', which opens Section 15—he has dropped in the course of Section 16.

24   *single essences* individual beings.

30   *bezo las manos* literally, kiss the hands: an expression of gratitude.

32   *the ram in the thicket* Genesis xxii. 13.

33   *Moses in the ark* Exodus ii. 3–10.

PAGE 21

2   *Joseph* Genesis xxxvii–xlviii.

4   *which pass awhile under the effects* which are accepted for a time as the effects.

7   *a miscarriage in the letter* The Gunpowder Plot was discovered
    after Francis Tresham, one of the conspirators, had sent a warning
    letter to Lord Monteagle advising him not to take his seat in
    Parliament. There is nothing to justify Browne's suggestion that
    this letter miscarried.

    victory of '88 the defeat of the Spanish Armada in 1588.

18 *the writing upon the wall* Daniel v. 5.

19-20 *the success of...Holland* the war of independence waged by
    the Dutch against their Spanish rulers ended in the truce of 1609
    after some thirty years of fighting.

    the Grand Seignieur the Turkish emperor. The anecdote has not
    been traced to its source.

29 *a revolution and vicissitude* a belief given current support and
    interest by Hakewill's *Apology*, first published in 1627. Sanna
    refers to Montaigne's essay 'Le profit de l'un est dommage de
    l'autre' (I, xxii) and, through Praz, to Dante, *Inferno* VII.

31 *intelligences* spiritual beings who were believed to move or guide
    the celestial orbs.

34-p. 22, l. 1 *an helix that still enlargeth* a continually extending
    spiral.

PAGE 22

4-6 *These must not...of nature* we should then speak of events as
    the outcome of chance only as we speak of the works of nature:
    both are in fact the work of God.

17 *that fools are only fortunate* that only fools enjoy good luck.

31 *wherein we resemble our Maker* in wisdom.

PAGE 23

5 *no just quarrel with nature for leaving us naked* 'que nous sommes
    le seul animal abandonné nud sur la terre nuë, lié, garrotté...là
    où toutes les autres creatures, nature les a revestuës de coquilles, de
    gousses, d'escorce, de poil, de laine' (Montaigne, *Essais*, II, 136).

8-9 *judicial astrology* the art of judging the occult influences of stars
    and planets upon human affairs; distinct from natural astrology,
    which calculates natural phenomena such as tides and eclipses.

11 *owe a knee* owe respect.

12–13   *ordered my indifferent...aspects* favoured me with the gift of good fortune, when birth left this undecided.

19   *Homer's chain* 'whose highest link, poets say prettily, is fastened to Jupiter's chair, and the lowest is riveted to every individual on earth' (Digby).

25   *each singular essence* every individual being or substance.

27   *bad construction* misinterpretation.

28   *these pair of second causes* fortune and providence.

32   *to compose these...angry dissensions* Browne's temperament inclines him consistently towards agreement and amity: compare his readiness to adapt himself to Roman Catholic ritual in section 3, and to withdraw any statement that conflicts with orthodox opinion in section 60.

PAGE 24

3   *that other* the triumvirate of Antonius, Octavius and Lepidus.

13   *too nearly* too intimately.

18   *demonstrating a naturality* showing how a supposedly miraculous event might occur naturally.

20   *Archidoxis* a book of philosophical or pseudo-scientific secrets, collected by Paracelsus. Browne, and not the devil, is the reader in question.

21–2   *brazen serpent* Numbers xxi. 9.

22   *worked by sympathy* those suffering from snake-bite were cured by looking at the brazen serpent set up by Moses: the kind of associative remedy that was common in occult medicine, and which Browne appears to have credited.

27   *miracle in Elias* otherwise Elijah; I Kings xviii. 35.

30, 32   *Sodom, Gomorrah* Genesis xix. 24–8.

34   *Josephus* a first-century Jewish historian, from whose *Antiquity of the Jews* Browne presumably took this statement.

PAGE 25

14   *Epicurus* Athenian philosopher of the third century B.C. He believed that the gods took no interest in the material world and human events.

17   *that fatal necessity of the Stoics* the Stoic philosophers, character-

ised by their austerity and indifference to pleasure and pain, believed fate to be superior to the gods, whose powers were confined to the material world.

18–19 *Those that...denied...Holy Ghost* the Macedonians, who maintained their heretical doctrine between 360 and 381.

24–5 *That villain...three impostors* identified by Sanna (p. 163) as Bonaventure des Périers, 1500–44, courtier to the Queen of Navarre, whose *De tribus mundi impostoribus*—a work declaring that mankind had been deceived by three impostors, Moses, Christ and Mahomet—was printed in 1537. The book was suppressed before it could be published, but it was offered for sale in the following year. The ensuing outcry forced des Périers to leave Paris.

27 *Machiavel* Nicolò Machiavelli of Florence, 1469–1527; notorious as the author of a political treatise *The Prince*, 1513, which advocated duplicity in political affairs.

28 *Lucian* second-century Greek writer of satirical dialogues.

PAGE 26

2–9 *a doctor in physic...poison of his error* these two recollections, with a third on p. 29, ll. 19–22, provide glimpses of Browne's intellectual activities during his residence abroad.

6–7 *three lines of Seneca* 'Post mortem nihil est, ipsaque Mors nihil. Mors individua est noxia corporis. Nec patiens animae... Toti morimur, nullaque pars manet Nostri' (marginal gloss). After death there is nothing; death itself is nothing. Death is the total destruction of the body: there is no question of a soul. We die wholly, and no part of us remains.

9–10 *relations of mariners* travellers' tales.

11 *Aelian, Pliny* Aelian's *Varia Historia* and Pliny's *Historia Naturalis* are largely compilations of other men's works uncritically repeated.

15 *Gargantua* Gargantua et Pantagruel, by Rabelais; first published about the middle of the sixteenth century. Urquhart's incomplete English translation appeared in 1653.

Bevis Sir Bevis of Hampton; a long rambling tale of knight-errantry first printed in 1500, with many subsequent editions.

17–18 *carry the buckler unto Samson* be worthy of comparison with him.

26–7 *atoms in divinity* two ancient philosophers, Leucippus and Democritus, had asserted that the universe was made of infinitely tiny particles of matter, which they called atoms. Browne appears to be saying that none of the fantastic suggestions that can be raised in divinity is quite as ridiculous as this scientific theory, which contradicts the account of creation given in Genesis.

PAGE 27

5 *whether Adam...hermaphrodite* the sense requires 'I dispute not', from line 1 above.

6 *the letter of the text* 'Let us make man'; understanding mankind, and not the male sex alone, to be meant.

11 *whatsoever sign the sun possesseth* through whichever of the twelve zodiacal signs the sun happens to be passing—that is, at whatever time of the year.

19 *Pantagruel's library* for the complete catalogue see *Gargantua et Pantagruel* II, 7. The library of St Victor contained, besides the treatise on evacuation which Browne mentions, nonsensical books on the geography of Purgatory, the cuckold at court, and the account of an apparition which appeared to a nun in labour.

22 *so serious a mystery* the operation of fortune, Browne's topic in Section 18, from which he has digressed to consider motives of disbelief.

23 *called to the bar* formally interrogated.

25 *Deucalion* a figure of Greek mythology analogous to Noah.

26–7 *seems not to me so great...always* since mankind is still wicked, as in the days of Noah. But this begs the question whether there had been a general flood, whether for that or any other reason.

30 *three hundred cubits* about 450 feet; the length of the Ark.

33 *the honest Father* 'This honest Father was St Augustine, who delivers his opinion that it might be miraculously done lib. 16 *de Civ. Dei* cap. 7' (Digby). Healey translates: 'We cannot deny that the angels by God's command might carry them thither.'

RELIGIO MEDICI

2 *planted by men* colonised.

7 *came over* to America after being released from the Ark.

8 *this triple continent* Europe, Africa and Asia.

10 *Ararat* on which the Ark grounded.

11 *make the deluge particular* say that the Flood was a local disaster.

14 *whereby I can make it probable* 'The connection of this passage
with the preceding one is not clear' (Sanna). Browne is arguing
against those who support their disbelief in a general flood by
contending that the subsequent period has been too short for
repopulating the world. He points out more than twice as long
a period has elapsed since the Flood than between the Creation
and the time of Noah.

15–16 *fifteen hundred years* the period between the Creation and the
Flood, determined by adding together the ages of the Patriarchs.

16–17 *four thousand years since* the period between the Flood and
Browne's own lifetime.

20 *'Tis a postulate* an unproved proposition.

23–4 *that Judas perished* the accounts of his death in Matthew xxvii.
3–5 and Acts i. 16–18 are contradictory.

25–6 *a doubtful word* Sanna points out, p. 168, that the word ἀπήγ-
ξατο in St Matthew's gospel means not only hanging but suffo-
cation.

31 *another intention* 'Let us make us a name, lest we be scattered
abroad upon the face of the whole earth' (Genesis xi. 4).

1–2 *under favour* subject to correction.

9 *Sarah* see Genesis xvii. 12; I Peter iii. 6.

15 *the opinion of tutelary angels* belief in guardian angels.

15–16 *when Peter knocked at the door* Acts xii. 13–15.

20 *that answered upon* who was replying to discussion about.

29 *Ptolemy* 'The king of Egypt who, according to the commonly
received tradition, caused the Jewish Scriptures to be translated
into Greek, and placed them in his newly established library at
Alexandria' (Greenhill).

30 *Alcoran* the Koran; properly, of the Moslems.

COMMENTARY AND NOTES

PAGE 30

3   *this without a blow* the Bible.

5   *Philo* a Jewish philosopher and writer of the first century.

10 *Zoroaster* or Zarathustra; a religious teacher of the sixth century B.C.

14 *this only* the Bible.

15 *those general flames* at the end of the world, when the whole of material creation will be burnt. See p. 52, l. 15: 'the last and general fever', and p. 55, l. 5: 'Now what fire.'

16 *confess their ashes* admit their impermanence.

17–18 *lost lines of Cicero* forty-eight of Cicero's orations are lost.

18–19 *combustion...Alexandria* probably several hundred thousand works were lost when the library was burnt at the siege of Alexandria during Caesar's campaign in Egypt.

21 *the Vatican* the papal palace at Rome, which contains a great library.

24 *Enoch's Pillars* 'Josephus says that the descendants of Seth erected two pillars, on which were engraven all the inventions and discoveries then known to mankind' (Greenhill). One was made to resist water, the other fire.

26 *Pineda* 'Pineda in his *Monarchia Ecclesiastica* quotes one thousand and forty authors' (marginal gloss).

28 *three great inventions of Germany* printing and gunpowder are certainly intended; the third is open to question. A marginal note in two MS. copies of *Religio Medici* assumes that the mariner's compass is to be included: Digby supposed that clocks were meant. See *Pseud. Epid.* v, 18: 'Polydore Vergil discoursing of new inventions whereof the authors are not known, makes instance in clocks and guns.'

PAGE 31

1   *at first* before the invention of printing, when books were copied by hand.

11 *the issue of Jacob* the Jews.

12–13 *the idolatry of their neighbours* under the Maccabean rulers and Herod. See I Maccabees i. 13–15.

16 *of conversion* to the Christian faith.

113

17 *obstinacy...constancy in a good* Browne does not share Montaigne's more sceptical appreciation that such terms are decided by one's point of view.

23-4 *from the name...Prophet* reject Christ in favour of Mahomet.

25 *it is the promise of Christ* John x. 16.

27-8 *four members of religion* Christian, Jew, Mahometan and pagan.

28 *we hold a slender proportion* Christians are outnumbered by those of other faiths.

PAGE 32

9-10 *unhappy method of angry devotions* regrettable practice of bigoted faith.

15 *fetched from the field* examples taken from warfare.

19 *which Aristotle requires* *Ethics*, III, 6-9.

22 *that easy and active part* in warfare.

24 *these have surpassed* martyrs have outgone soldiers in courage.

PAGE 33

1 *The Council of Constance* 1414-18, designed to end the Great Schism. Besides condemning Huss as a heretic, the Council ordered that Wycliffe's body should be removed from consecrated ground.

1-2 *John Huss* a Bohemian reformer, 1369-1415. As a Wycliffite he attracted the enmity of the Pope, was excommunicated and finally burnt at the stake.

8 *Socrates* put to death 399 B.C. on a charge of corrupting young men and of not respecting the state religion. The mistaken idea repeated by Browne seems to have originated with the Humanists. One of the *Colloquia* of Erasmus speaks of Socrates as a man whose desire to obey divine will reveals a truly Christian spirit.

9-10 *the miserable bishop* 'Vergilius, Bishop of Salzburg in the eighth century, said...to have been burnt for heretically asserting the existence of Antipodes' (Greenhill). St Vergilius was a gifted mathematician and scientist. He was accused of heresy in consequence of his advanced beliefs, probably before his consecration as bishop, but evidently the charge was rejected.

21 *leaven and ferment* an agency that produces profound change by working inwardly.

33 *miracles in the Indies* attributed to Giuseppe Acosta, 1539–1600, to whom Burton refers in *The Anatomy of Melancholy*, III, 4. For a more detailed note, see Sanna, p. 180.

PAGE 34

3–4 *transmutation...our Saviour* during the celebration of Mass.

13 *against...above...before nature* against natural law, above the power of nature, or before any natural world existed.

15 *too narrowly define the power of God* 'Je ne trouve pas bon d'enfermer ainsi la puissance divine soubs les loix de nostre parolle' (Montaigne, II, 223).

19 *Esdras* see II Esdras iv. 5.

20 *pose mortality* ask man to perform.

29–30 *fathered on the dead* attributed to the power of dead saints.

33 *the cross that Helena found* St Helena, mother of the Emperor Constantine, visited the Holy Land in 326 and is reputed to have discovered the cross of the Crucifixion.

PAGE 35

1–2 *I excuse not Constantine* I refuse to believe that Constantine was miraculously protected.

5 *piae fraudes* pious frauds.
    nor many degrees before nor much better than.

6 *Baldwin* crowned King of Jerusalem on Christmas Day 1100 at the conclusion of the First Crusade.

10 *speak naturally...not salve the doubt* discuss miracles as though they were natural events and do not settle the mystery.

15 *the Ancient of Days* Jehovah; see Daniel vii. 9.

21 *the cessation of oracles* compare *Pseud. Epid.* VII, 12, 'That oracles ceased or grew mute at the coming of Christ', and Milton, *Hymn on the Morning of Christ's Nativity*, 145–52.

23 *as Plutarch allegeth for it* 'That the oracles were nourished by exhalations drawn up from the earth, and that when these were exhausted the oracles famished and died for want of their accustomed sustenance' (Digby).

24 *that supernatural solstice* See p. 18, l. 29, *its supernatural station*, and note.

26 *at his death* During the Crucifixion 'there was darkness over the whole land until the ninth hour'.

27 *the devil himself* 'in his oracle to Augustus' (marginal gloss). The oracle is quoted in *Pseud. Epid.* VII, 12 with Browne's translation, which begins:

An Hebrew child, a god all gods excelling,
To hell again commands me from this dwelling.

31 *Megasthenes* Greek ambassador of about 300 B.C. who wrote a work on the religion and customs of India. *Herodotus* Greek historian of the fifth century B.C.

33 *Justin* a third-century Roman whose condensation of the universal history of Pompeius Trogus was popular with medieval readers.

PAGE 36

8 *his death also* probably alluding to Philo, 'who expressly says that Moses wrote the account of his death and burial prophetic-ally' (Greenhill).

10–11 *doubtful conceit of spirits* incredulity towards their existence.

12–13 *ladder and scale of creatures* See Browne's own description on p. 39, ll. 19 ff., 'There is in this universe a stair...'.

14–15 *that there are witches* Browne's belief in witches persisted at least until 1664, when he gave evidence at the trial of Amy Duny and Rose Cullender, who were convicted of witchcraft and executed.

27–8 *use with men...carnality* perform acts of sexual intercourse, as human beings do.

PAGE 37

6 *the maid of Germany* 'That lived without meat upon the smell of a rose' (marginal gloss). The maid was probably Eva Flegen of the county of Mörs, who after a serious illness in 1597 gradually gave up eating, affirming that she was miraculously fed every third day at sunrise 'by a honey-like dew which very much refreshed her'. She was reported as still fasting in 1625, but by

1628 she had been exposed as a fraud and imprisoned. See correspondence in *T.L.S.* 24 July and 21 August 1948.

13–14 *actives, passives* creative forces, receptive instruments.

15 *philosophy...witchcraft* some scientific knowledge is derived from occult lore, e.g. herbal remedies.

21 *Paracelsus* a Swiss physician, 1493–1541. One of the earliest empirical philosophers, he was also a Neoplatonist, holding that we know God only in so far as we are God.

22–3 *Ascendens...opera Dei* a constellation as it rises discloses many things to those who investigate the mighty works of nature. 'Thereby is meant our good angel appointed us from our nativity' (marginal gloss).

PAGE 38

2 *Hermetical philosophers* adherents of the occult philosophical doctrines attributed to Hermes Trismegistus: see note on p. 12, l. 7, *Hermes*.

6 *makes no part of us* is not a constituent part of our human nature.

8 *radical heat* the heat that is a necessary condition of life. See section 43, pp. 51–2.

19 *Cancer's back* the zodiacal sign of the Crab, which the sun enters on 21 June.

PAGE 39

11 *an old one of Plato* expressed in his *Phaedo*, a work on immortality.

17–18 *I confess them...like that of God* I admit that my opinions are very superficial; that generally I define angels in an indefinite way as being like God.

19 *between ourselves and fellow-creatures* probably 'and God' should be supplied after this phrase.

22 *creatures of mere existence* such as stones, without power of movement or sensation.

26 *there should be yet a greater* a typical example of the inductive method of the Schools, which empirical philosophy was beginning to discredit.

27–8 *the definition of Porphyry* 'essentia rationalis immortalis' (marginal gloss): an essence rational and immortal. Porphyry, a

Neoplatonist of the third century, was author of a long treatise against Christianity. See p. 25, l. 31, 'I confess I have perused them all'.

PAGE 40

2–3 *upon the first motion...study or deliberation* that they possess knowledge instantaneously, without any preliminaries of reasoning or reflection.

4–5 *define by specifical differences...accidents and properties* determine by innate characteristics of species what we identify through external appearances and attributes. Compare p. 78, ll. 29–30.

7–8 *specifical but the numerical forms of individuals* both the species and the individual characters of things.

9–10 *by what reserved difference...numerical self* by means of what special peculiarity each separate entity acquires its individual nature, as well as its natural kind.

12 *a faculty...inform none* a power capable of moving a body without entering into it.

12–13 *ours upon restraint* our souls bound by limitations.

14 *Habakkuk* Apocrypha, Bel and the Dragon 33–9. The two woodcuts after Holbein, reproduced on p. 41, illustrate the story of Habakkuk; the first showing him about to be snatched up as he goes to feed the reapers, the second his being deposited in the lions' den.

  Philip Acts viii. 40. The assumption that Philip was miraculously transported to Azotus rests upon a misunderstanding of the text.

23 *at the conversion...rejoice* Luke xv. 10.

24 *that great Father* St Augustine, in his *Confessions*, XIII, 3.

25 *Fiat lux* let there be light.

27–8 *we style...accident* we call light a mere appearance.

29 *light invisible* 'Whensoever any affirmation is false, the two names of which it is composed, put together and made one, signify nothing at all. For example, the word *round quadrangle* signifies nothing, but is a mere sound' (Hobbes, *Leviathan*, 1651). Hobbes's stricture shows how far Browne was from sharing the outlook of the new philosophers in his fondness for paradox.

33 *actually existing what we are but in hopes* the angels enjoy in

actuality the kind of existence we can only aspire to know in heaven.

1 *that amphibious piece* Browne keeps the idea in hand and expands it later in the same paragraph.

3 *makes good the method of God and nature* proves by demonstration the systematic purpose behind natural and divine creation.

4–5 *incompatible distances* between bestial and angelic.

7 *upon record of holy Scripture* Genesis ii. 7.

8–9 *a pleasant trope of rhetoric* a figurative expression.

10–11 *a rude mass* in the early stages of foetal growth.

11 *creatures which only are* having existence without sense, as stones.

21–2 *Moses...description* Browne assuming that Moses was the author of Genesis, which says nothing about the creation of an invisible world.

27–9 *perhaps the mystical method...Egyptians* expressing mystical truths by means of allegory; a technique which Moses learned from the Egyptians, who 'were nearer addicted to those abstruse and mystical sciences' (p. 74, ll. 16–17). See also p. 18, ll. 32–3, and p. 60, ll. 6–7.

31 *the first moveable* the outermost of the eight or nine spheres surrounding and enclosing the earth, according to Ptolemaic astronomy.

33 *extremest circumference* referring to the same conception of a geocentric universe. Browne admits his scepticism towards the Copernican theory on p. 94, l. 3 and on p. 96, l. 11.

1 *extract from the corpulency of bodies* separate existence from its material forms.

2 *first matter* uncreated chaos, matter without form: an Aristotelian concept defined in 1687 as 'mere possibility of being'.

8 *at a distance even, in himself* contained by God even when at a distance from him: a confusion of thought, since if the angels are inseparable from the essence of God, they cannot be 'at a distance' from him.

17-18 *it was necessary...this homage* the final cause of man's creation: another typically Scholastic inference.

22 *sworn he would not destroy it* in the covenant after the Flood; Genesis ix. 11.

24-5 *as weakly that the world was eternal* Aristotle provides no convincing proof that the world is eternal.

27 *Moses hath decided that question* Browne again assuming that Moses was the author of Genesis.

28 *new term of a creation* new definition of creation: that provided in Genesis, which describes God calling everything into existence from a state of darkness and void.

33-4 *generation not only...but also creation* creation and generation are contradictory terms, since they involve nothing becoming something.

34-p. 44, l. 3 *God, being all...an essence* Browne treats his paradox as though it were a logical proposition, not noticing that the conclusion he draws from it depends upon his manipulating the sense of undefined terms to suit his purpose.

PAGE 44

7 *the text* that of Genesis: 'and the Lord God formed man of the dust of the ground'.

8 *played the sensible operator* moulded man from a material substance. Here Browne disagrees with St Augustine: 'We are not to conceive this carnally, as we see an artificer work up anything into the shape of a man by art' (Healey, XII, 23).

9 *make him* to shape him out of raw materials.

  separated the materials 'And out of the ground the Lord God formed every beast of the field, and every fowl of the air' (Genesis ii. 19). This account of Creation differs from that given in Genesis i. 24, to which Browne refers in his previous sentence, 'At the blast of his mouth were the rest of the creatures made'. Compare p. 58, l. 3: 'a separation of that confused mass into its materials', and again four lines later.

12 *harder creation* the idea conflicts with Browne's assertion on p. 34, ll. 10-11 that to God 'all things are of an equal facility'.

13 *these two affections* the attributes of immortality and incorrupti-

bility, which Plato had affirmed in his *Phaedo* and which Aristotle had not denied.

16 *concerning its production, much disputed* The Traducianists asserted that the soul was not infused into the growing foetus by God but transmitted to the child by its parents at the moment of conception. The doctrine was supported by the Lutherans, and contested by the Calvinists.

19 *Paracelsus* see note under p. 37, l. 21. 'Man, whom Paracelsus would have undertaken to have made in a limbeck, in a furnace' (Donne, Sermon VII).

21 *deny traduction* oppose the Traducianist doctrine. Browne appears to favour it: see below, ll. 31 ff.: 'For if the soul....'

23 *Augustine* St Augustine of Hippo, 354–430; one of the great theologians of the early Christian Church, to whom Browne is indebted for several ideas.

23–4 *creando...creatur* by the act of creating, grace is poured on the world: the pouring of grace is itself a creative act. The author of the remark was in fact Peter Lombard, 1100–60, the so-called 'Master of Sentences' whose *Sententiarum libri quatuor* was written about 1150.

PAGE 45

2 *improper organs* monstrous bodies.

2–4 *the soul...inorganical* the inorganical powers of the soul were held to be reason, 'which understandeth and judgeth', and will, 'whereby man electeth or refuseth whatsoever intelligence or reason hath judged to be good or evil'. The bodily perceptions were required so that the soul could be acquainted with her earthly surroundings. Without what Browne calls 'a proper disposition of organs', her information would be unreliable.

6 *crasis* 'that is, a temperature of the whole body, or a kind of elemental matter, moving and stirring the body' (John Woolton, *A Treatise of the Immortality of the Soul*, 1576).

9 *and that the hand of reason* the senses are the agency by which reason acquaints itself with the material world.

10–11 *such as reduced...divinity* probably referring to *De usu partium*, III, x, in which Galen exclaims at the wisdom of man's Creator.

19–22 *Thus we are men…in us* a curious change of style from high-flown philosophical language to colloquial simplicity.

25 *elemental composition* something made of the material elements of earth, water, air and fire.

PAGE 46

6–7 *that of Nebuchadnezzar* 'And his body was wet with the dew of heaven, till his hairs were grown like eagles' feathers, and his nails like birds' claws' (Daniel iv. 33).

24 *his victory in Adam* whose Fall brought death into the world.

26 *quid fecisti?* What has thou done? Genesis iii. 13.

PAGE 47

28 *this dilemma* a choice between two equally unfavourable alternatives.

PAGE 48

5 *in our chaos* as a spirit for whom no body has yet been formed.

6 *within the bosom of our causes* Browne follows Aquinas in believing that each soul is newly created by God—its first cause—and infused into the body destined for it.

10–11 *ourselves being not yet without life* Sanna follows the reading of the Pembroke College MS., 'being yet not', and suggests that a word has been lost after 'being'; perhaps 'imperfect': unnecessary if 'being' means having existence. The sense of the argument is clear: within the womb we seem to possess only a vegetative soul, but our higher faculties already exist, waiting until they can be exercised through contact with the external world.

12 *opportunity of objects* occasion of encountering things upon which thought and reason can be engaged.

13 *soul of vegetation* see note p. 14, l. 9, 'three distinct souls'; here that of growth.

19 *that ineffable place of Paul* II Corinthians xii. 4.

20 *philosopher's stone* a supposed substance with the power of transmuting other metals into gold or silver; also able to cure all wounds and diseases. Browne seems to regard the stone as gold in the highest state of refinement.

26   *observed in silkworms* the metamorphosis of caterpillars.

    turned my philosophy into divinity led me from natural history to religious contemplation. Here, as in other passages of *Religio Medici*, Browne uses the term 'philosophy' to mean natural philosophy, or science.

33–4 *est mutatio...microcosmi* there is a final transformation by which that noble extract of the microcosm is brought to perfection.

33–p. 49, l. 1 *a natural and experimental way* in the light of general experience. Browne was not an empiricist, and this older sense of the term 'experimental' was the more familiar one during his lifetime. The earliest example of the term 'experimental philosophy' quoted by the *O.E.D.* is found in a work published in 1651.

PAGE 49

1   *but a digestion* merely a stage in the process of extracting the essence of matter.

13   *in a tempest* probably a reference to one of Browne's voyages. He was shipwrecked while returning from Ireland in 1630.

16   *Quantum mutatus ab illo* how much changed from his former self; referring to the shade of Hector whom Aeneas encountered in Hades, *Aeneid*, II, 274.

22   *upon the courage...issue* in the assurance of having children.

25   *counterfeit subsisting in our progenies* illusory living on in our children.

PAGE 50

1–2   *testament of Diogenes* 'who willed his friends not to bury him, but to hang him up with a staff in his hand to fright away the crows' (marginal gloss).

3   *Lucan* Roman poet and Stoic, 39–65, who joined a conspiracy against Nero and was compelled to commit suicide.

10   *temper of crows and daws* constitution of long-lived creatures.

11   *before the Flood* the patriarchs are credited with lives of several centuries: Noah, for instance, died at the age of 950 (Genesis ix. 29).

12   *I may outlive a jubilee* he did; dying in 1682 at the age of 77.

13   *one revolution of Saturn* thirty years.

    thirty years Browne was born in October 1605.

17   *and begin to be weary of the sun* like Macbeth, at v, v, 49. What
    follows three lines later may be another recollection of the play.

23   *I use myself* accustom myself. Browne suggests that he had
    adopted Stoicism as his private philosophy.

25   *not by physic* although himself a physician.

PAGE 51

13   *pretence unto excuse* claim to be excused.

15   *succeeds in time* is prolonged or extended.

17   *figures in arithmetic* Browne is thinking of indices or exponents.

21   *upon Cicero's ground* 'Neque me vixisse poenitet; quoniam ita
    vixi, ut non frustra me natum existimem.—*De Senectute* 23'
    (Digby). 'I'm not sorry that I have lived, because I have lived in
    such a way that I do not think I was born in vain.'

29   *Aeson's bath* a magic device to renew the youth of an old man.

33   *an able temper* a balanced constitution.

PAGE 52

 4   *found themselves* base their opinions.

    vital sulphur The alchemists took sulphur to be one of the
    ultimate elements of all material substances. The vital sulphur or
    radical balsam mentioned here was believed to act as a healthful
    preservative within the body.

 7   *to determine them* to decide how long they should be.

13–14 *must not expect the duration...constitution* must not expect to
    last as long as its fabric would naturally endure. Compare p. 55,
    ll. 2–3.

16   *before six thousand* six thousand years after the Creation; the
    supposed term of the world's existence. Compare p. 55, ll. 24–5.

PAGE 53

 1   *Zeno* first of the Stoic philosophers, who lived 335–263 B.C.

 4   *his own assassin* in anticipation of 'his own executioner', p. 79,
    l. 10.

4–5 *suicide of Cato* who took his life in 46 B.C. to avoid falling into the hands of Julius Caesar, whose increasing despotic power he had opposed.

9 *valiant acts of Curtius* who, to close a chasm which had opened in the Roman Forum, leapt with his horse into the abyss. *Scaevola*, after failing to kill Porsenna, held his right hand in fire to prove his indifference to pain. *Codrus*, an early king of Athens, learning of a prophecy that the invaders would be victorious if his life were spared, disguised himself as a woodcutter, provoked a quarrel with a group of invading warriors and was killed by them.

10 *that one of Job* Job ii. 9–10.

10–11 *no torture to* no pain comparable to.

18–19 *I that have examined the parts of man* Browne expresses pardonable pride at having practised anatomy, probably at the medical schools of Montpellier and Padua, at a time when theological objections to this study were only slowly being overcome. In the year of Browne's birth Bacon had found 'much deficience' in the study of anatomy: 'for they enquire of the parts, and their substances, figures, and collocations; but they enquire not of the diversities of the parts, the secrecies of the passages, and the seats or nestling of the humours' (*Advancement of Learning*, II, x, 5). Here Browne seems abreast of current thought; but his continuing respect for Galen's *De usu partium* (p. 17, l. 22 and note) suggests a double allegiance; and his satisfaction at not locating the anatomical seat of the soul (p. 45, ll. 11–14) elevates theology at the expense of science.

21 *the thousand doors that lead to death* Compare:

> I know death hath ten thousand several doors
> For men to take their exits.
>
> (Webster, *The Duchess of Malfi*, IV, ii, 225–6.)

24–5 *the new inventions of death* newly invented lethal weapons. The musket and the dag, introduced during the sixteenth century, were still comparatively recent inventions.

30–1 *the misery of immortality...immortal* Christ, who was immortal though a man, did not undertake the misery of physical immortality.

PAGE 54

3 *in his own sense* from the point of view of his philosophy.

5 *we are in the power...death is in our own* Here Browne comes perilously close to 'that insolent paradox that a wise man is out of the reach of fortune' (p. 22, l. 18).

8 *mortification* deadening the bodily appetites by asceticism, or by inflicting pain or discomfort on the body.

12 *horae combustae* 'that time when the moon is in conjunction and obscured by the sun' (marginal gloss): meaning here the interval between birth and baptism.

16 *sensible affections of flesh* a body responsive to sensation.

21 *memento mori* remember that you have to die.

21-2 *Memento quatuor novissima* remember the four last things.

25 *Rhadamanth* in Greek mythology, one of the judges of the lower world.

27 *Sibyl* the prophetess whom on p. 60 Browne ungallantly describes as the devil of Delphos.

PAGE 55

1 *the world draws near its end* a commonplace of early seventeenth-century thought, which prompted the massive counter-argument of Hakewill's *Apology*. Browne repeats the idea with characteristic lack of misgiving, introducing it again on p. 56.

2-3 *upon the ruins of its own principles* by the exhaustion or cessation of its vital processes.

5-6 *what fire should be able to consume it* Compare:

> The Schoolmen teach that all this globe of earth
> Shall be consumed to ashes in a minute:
> ...'twere somewhat strange
> To see the waters burn: could I believe
> This might be true, I could believe as well
> There might be hell or heaven.

> (Ford, *'Tis Pity She's a Whore*, v, v.)

10-11 *but rather seem...the method and idea* this attitude conflicts with Browne's usually literal interpretation of events in Scripture, but is consistent with his fondness for symbols and types.

126

17 *a vulgar and illustrative way* allusively, in a form easily understood.

19–21 *according to different capacities...each single edification* suiting different mental powers, may conform with our religious faith without endangering the moral instruction which such passages impart.

24–5 *Elias' six thousand years* 'There is a report that in the books of Elias the prophet it was recorded that the world should last six thousand years: two thousand under vanity, unto Abraham; two thousand under the law, to Christ; and two thousand under Christ, to the Judgement' (Vives, commentary upon *The City of God*, xx, 8, edition of 1610).

30 *many melancholy heads* Sanna suggests Browne's general indebtedness to Burton's *Anatomy of Melancholy*, III, iv, 1 for this passage, and finds other parallels with Burton throughout *Religio Medici*.

PAGE 56

1 *old prophecies* 'In those days there shall come liars and false prophets' (marginal gloss). Matthew xxiv. 23–4.

8–10 *Antichrist...come these many years* Compare p. 7, ll. 7–8.

11 *those ridiculous anagrams* referring to speculation on the number of the Beast, Revelation xiii. 18. Hebrew and Greek had no numerals, but gave each letter of the alphabet a numerical value.

12 *philosopher's stone* here used figuratively, as an object of unavailing search.

16 *that the world draws near its end* acknowledging the private belief of p. 55, l. 1 to be general. ''Tis too late to be ambitious. The great mutations of the world are acted' (Carter, *Urn Burial*, p. 45).

19 *the saints under the altar* Revelation vi. 9–10.

22 *This is the day* Judgement Day.

25 *those seeming inequalities* described on p. 23, ll. 2–4.

30 *whose memory hath only power* the recollection of which alone has power.

PAGE 57

2 *Seneca* Roman moralist and tragic dramatist of the first century, whose formulation of Stoical outlook had a prolonged influence on

127   9-2

European thought. *Ipsa sui pretium virtus sibi*, three lines above, paraphrases an idea in his *De vita beata*, whose text Browne may have been unable to consult: see p. 2, ll. 3–4.

11 *natural inclination…unto virtue* showing Browne's sympathy with the optimistic assessment of man's nature, derived from Neoplatonism, which his remarks on p. 51, ll. 4–8 contradict.

12–14 *yet not in that resolved…forget her* but not with such constant resolution that I could not yield to the temptation to forget virtue.

18 *Euripides* Greek tragic poet of the fifth century B.C. He sympathised with the sceptical outlook of his age, and suggested that the morality of the gods was inferior to that of good men.

Julian Julian the Apostate, 332–63; Roman emperor from 361. His imperial policy was to degrade Christianity and to promote paganism, by means which included publishing treatises against Christian doctrine.

29 *a single experiment* a single test of experience.

PAGE 58

17 *accidentally and upon forfeit* Browne refers to the Fall as an event not designed by God, by which man forfeited the immortality conferred upon him at his creation.

24 *Let us speak* putting this.

32 *a sensible artist* a perceptive philosopher or man of science.

34 *that devouring element* fire.

PAGE 59

1 *made good by experience* experimentally proved. 'Some lying boast of Paracelsus, which the good Sir Thomas Browne has swallowed for a fact' (Coleridge, quoted by Greenhill). But if so, Browne was capable of gulling others with the same fable. In February 1648 the microscopist Henry Power wrote to Browne for details about the 're-individualling of an incinerated plant…hoping to find as much willingness to communicate as ability to evince the certainty of this secret' (Keynes, pp. 280–1). Browne's reply is not preserved.

5 *more perfect and sensible structures* the bodies of men.

8 *as Ezekiel* Ezekiel xxxvii. 5–10.

13–14 *that elegant Apostle* St John, in his Revelation; 'elegant' in respect of his literary style.

PAGE 60

1 *tenth sphere* see note under p. 46, *the first moveable*, and below, l. 19. The tenth sphere, or Empyrean, lay beyond the moving spheres of the moon, the sun, the five planets, the fixed stars and the Primum Mobile.

2 *this sensible world* material creation, palpable to sense.

9 *he desired to see God* Exodus xxxiii. 18.

. *petitioned his Maker...contradiction* asked God to contradict himself.

12–13 *Dives...Lazarus* Luke xvi. 19–31.

16 *our glorified eyes* the power of sight we shall possess in heaven.

17–18 *the visible species...intellectual* that vision will be as unlimited in heaven as thought is on earth, so that hell will be visible from heaven although remotely distant.

19 *beyond the tenth sphere* outside the Schoolmen's universe.

20 *in a vacuity* Aristotle supposed that the visible emanations of material objects were carried on the air, and that consequently sight could not pierce completely empty space such as he conceived to exist beyond the Empyrean.

22 *the visible rays of the object* material objects were believed to project images of themselves upon the eye.

24 *prepare and dispose* impress itself upon the receptive medium which carries it.

30–1 *flame that can...purify the substance of a soul* a characteristic example of Browne's pseudo-scientific method; examining a spiritual concept from the point of view of commonsense materialism which he assumes for the sake of the paradox.

PAGE 61

5–7 *in this material world...powerfullest flames* 'There are some creatures that are indeed corruptible, because mortal, and yet do live untouched in the midst of the fire' (Healey, XXI, 2). See also XXI, 10: 'whether the fire of hell, if it be corporeal, can take effect upon the incorporeal devils'.

129

21-2 *fire, water, earth and air* the four Aristotelian elements, of which the material creation was supposedly constituted.

PAGE 62

3 *its dilated substance* the macrocosm.

7 *in posse* the state of potential being, as opposed to that of *in esse*, or actual existence.

10-11 *that little compendium of the sixth day* man; as on p. 18, l. 21.

22 *those flaming mountains* volcanoes.

25 *Lucifer keeps his court in my breast* a statement incompatible with p. 57, l. 11, and with p. 90, l. 14.

   Legion Mark v. 9.

26 *Anaxagoras* Probably Browne meant either Leucippus, a Greek philosopher of the fifth century B.C., who believed that an infinite number of worlds had been created from an initial vortex, or Epicurus, who held the same belief two centuries later. See Healey, XI, 5, 'Epicurus' dream of innumerable worlds'.

27 *Magdalene* St Mary Magdalen, from whom Christ cast out seven devils; see Luke viii. 2.

PAGE 63

5 *never afraid of hell* the buoyant confidence of Browne's outlook is not always an advantage to him. Here, as in the opening sentence of Part II, Section 8 (p. 84), he seems complacent rather than spiritually serene.

8-9 *to be deprived of them is perfect hell*

   Think'st thou that I, who saw the face of God,
   And tasted the eternal joys of heaven,
   Am not tormented with ten thousand hells,
   In being deprived of everlasting bliss?

          (Marlowe, *Tragical History of Dr Faustus*, 312-16.)

18-19 *they go the surest way to heaven...hell* Browne is picking up the Senecan ideas he had discussed on p. 57, ll. 6ff.

19-20 *other mercenaries* Sanna, p. 225, sees an attack on the Jesuits; 'mercenaries' in the sense that their good works are performed for the private end of avoiding damnation.

25 *abyss and mass* a contradictory doublet; presumably an amount huge enough to fill a gulf.

4 *poetical fictions converted into verities* e.g. for those who wrote about a fictitious Hades to become acquainted with the reality of Hell.

18-19 *in their own kind* according to their own nature: from which, as Browne admits in line 9 above, man has lapsed.

23-4 *these great examples of virtue* the pre-Christian philosophers, patriarchs and others.

31 *Phalaris' bull* a bronze figure inside which criminals were roasted alive at the order of a Sicilian tyrant Phalaris.

15-16 *whilst we lie...vice* while we are guarding ourselves against attacks from one vice.

20 *it is a lesson to be good* we cannot be virtuous without study; again reversing the idea of man's 'natural inclination unto virtue' (p. 57, l. 11).

21 *by the book* according to rule or custom.

28 *irregular humour* unpredictable temper or whim.

In brief, we are all monsters 'Or, à ce conte, aux plus avisez et aux plus habilles tout sera donc monstrueux' (Montaigne, II, 222).

29 *a composition of man and beast* apart from the inconsistency of thought with p. 16, l. 1, an example of untidy thinking; since the concept 'man' cannot include other concepts distinct from man.

30 *Chiron* a creature half man, half horse, who was tutor to Achilles.

31-2 *to sit but at the feet* to remain subservient.

5 *Strabo* Greek geographer of the first century B.C. who compared the known world to a cloak.

11 *our reformed judgement* to Protestant opinion.

23 *Sectaries* English Protestant dissenters.

24   *Atomist* materialist; following the scientific philosophy of
Democritus and Epicurus that all matter was composed of
atoms, and that the human soul was dissolved at death.

Familist member of the Family of Love; a religious sect introduced to England from Holland during the reign of Elizabeth,
who issued a proclamation against them in 1580.

PAGE 68

10–11   *sentence Solomon* Romans ii. 12.
12–13   *letter and written law of God* Romans iii. 23.
19–20   *the eye of this needle* Matthew xix. 24.
21   *little flock* Luke xii. 32.
25   *anarchy in heaven* lack of order and spiritual degree among the
blessed.

hierarchies amongst the angels three divisions, each comprising
three orders: seraphim, cherubim, thrones; dominions, virtues,
powers; principalities, archangels, angels. According to Aquinas,
only the last two have any concern with human affairs.
29   *bring up the rear* be the last man to enter.

PAGE 69

14   *decree of that synod* predestination.
15–16   *the world was before the Creation...had a beginning* repeating
the paradox already expressed on p. 13, ll. 24–7.
20–1   *that do decry...upon faith* alluding to the Calvinists, whose
doctrine includes justification by faith without works.
21   *take not away merits* do not, by their affirmation, deny the
efficacy of good works.
22   *enforce the condition of God* compel God to act in accordance with
his declaration. Browne may be referring to Genesis xv. 6 or,
more probably, to Romans i. 17.
26   *Midianites* Judges vii. 5–7.

PAGE 71

5–6   *if I hold...myself* if I report myself truly.
16–17   *at the sight of a toad...destroy them* possibly influenced by
St Augustine: 'nothing at all in nature being evil...but every-

thing from earth to heaven ascending in a degree of goodness'
(Healey, XI, 22).

21 *in balance with* agreeing with.

23 *the eighth climate* the eighth of the regions of the earth, of which
there were twenty-four between the equator and the two polar
circles.

27 *I have been shipwrecked* on his return voyage from Ireland in 1630.

PAGE 72

1 *come to composition* make a treaty or compact.

5 *numerous piece of monstrosity* the many-headed monster of Horace,
Epist. I, i, 76, 'Belua multorum es capitum'; imitated by Sidney
in *Arcadia*, II, 'O weak trust of the many-headed multitude'; by
Daniel in *Civil Wars*, II, xii, 'This many-headed monster multitude', and by Shakespeare in *Coriolanus*, IV, i, 2. Browne appears
to be striking a familiar attitude rather than expressing a genuine
contempt. He has only a little earlier (p. 66, l. 28) confessed
to being a monster himself.

8 *Hydra* a fabulous many-headed snake, killed by Hercules.

17 *casting of account* reckoning up, balancing books.

17-18 *three or four men together...below them* 'Thus, 1,965. But why
is the 1 said to be placed below the 965?' (Coleridge, *Literary
Remains*). The question remains unanswered.

21 *place them below their feet* subordinate poor but worthy men to
rich upstarts.

25-6 *the bias of present practice* Under James I the purchasing of
knighthood and other honours had become notoriously common.

30-1 *everyone having a liberty...riches* alluding to the growing
wealth and importance of the merchant classes.

32 *general and indifferent* expansive and impartial.

PAGE 73

4-6 *yet if we are directed...our reasons* 'attenendosi all'insegnamento
di S. Tommaso, il Browne fa qui una netta distinzione fra le
virtù naturali e le virtù teologali' (Sanna, p. 236). The theological
virtues are faith, hope and charity.

12 *rhetoric...miseries* tale of misfortune.

24 *and truly, I have observed* the beginning of a digression on physiognomy, followed by another on the uniqueness of things,
after which Browne returns to the subject of charity at the
beginning of Section 3.

PAGE 74

1 *signs and bushes* a bush hung above a door was the sign of a
vintner.

7–8 *Adam assigned...its nature* Genesis ii. 19.

13 *in mine own hand* the lines in his palm.

18 *vagabond...Egyptians* gipsies.

21–2 *so many millions of faces...none alike* 'Nothing more variable
than faces and countenances: yet men can bear in memory
infinite distinctions of them; nay, a painter with a few shells of
colours, and the benefit of his eye, and the habit of his imagination, can imitate them all that ever have been, are, or may be, if
they were brought before him' (Bacon, *Advancement of Learning*,
II, x, 2).

23 *how there should be any* how there could be any.

32–3 *pattern or example of everything* According to Platonic belief,
every object is modelled upon an Idea or essence existing in the
mind of God.

PAGE 75

1 *herein it is wide* astray, defective; since, following the Platonic
argument, no picture could represent the Idea from which the
individual members of any species are derived.

15 *so narrow a conceit* so limited a conception.

16 *is only to be charitable* is the only real form of charity.

26–7 *wear our liveries* clothe their minds in our thoughts.

30 *this part of goodness* disseminating knowledge.

PAGE 76

9 *my acquired parts* his learning.

22 *parenthesis on the party* digression on a minor point.

27 Βατραχομυομαχία the title of a mock-epic poem falsely ascribed
to Homer, generally translated 'The Battle of the Frogs and the
Mice'.

28 σ *and* τ *in Lucian* 'His *Judicium Vocalium* is an amusing speech by Sigma before the Vowels (the judges in a mock trial) complaining of Tau for interfering with the other consonants' (Greenhill).

29 *the genitive case in Jupiter* 'whether Jovis or Jupiteris' (marginal gloss).

How many synods 'Combien de querelles et combien importantes a produit au monde le doubte du sens de cette syllabe: *hoc*!' (Montaigne, II, 223).

31 *Propria quae maribus* 'Perhaps most of the readers of this book will require to be informed that "Propria quae maribus" is the beginning of some (formerly) well-known lines of the old Latin grammar' (Greenhill).

32 *Priscian* famous Roman grammarian of the sixth century. The expression 'to break Priscian's head' was proverbially applied to those who spoke false Latin.

Si foret...Democritus If Democritus were on earth, he would laugh at it. Horace, *Epist.* II, i, 194.

PAGE 77

3 *Actius' razor* 'Actius Navius was chief augur who...admonishing Tarq. Priscus that he should not undertake any action of moment without first consulting the augur, the king demanded of him whether by the rules of his skill, what he had conceived in his mind might be done: to whom when Actius had answered it might be done, he bid him take a whetstone which he had in his hand, and cut it in two with a razor; which accordingly the augur did' (Digby).

9–10 *these are the men* scholars.

10–11 *when they have...their exits* great men.

11 *their scenes* lives of great men.

14–15 *no reproach to the scandal of a story* a historian is never reproached for injuring reputations. See *Pseud. Epid.*, ed. 1658, index: 'Stories indiscreetly transmitted, what mischief they do to posterity.'

19–20 *mysteries and conditions* trades and social classes: compare 'trade and mystery of typographers', p. 31, l. 4.

23–6 *le mutin anglais...ivrogne* rebellious English, swaggering

Scotch, unnatural Italian, mad Frenchman, cowardly Roman, thieving Gascon, arrogant Spaniard, drunken German.

27 *calls the Cretans liars* Titus i. 12.

28–9 *bloody...as Nero's was* Nero, Roman emperor 54–68, had his mother poisoned when his attempt to murder her by drowning had failed.

PAGE 78

1 *Democritus* 'the laughing philosopher' of the fourth century B.C. whose name Burton assumes in *The Anatomy of Melancholy*.

3 *Heraclitus* philosopher and scientist of about 500 B.C.
    moves not my spleen doesn't make me laugh.

9–10 *thus virtue...an idea* if vice did not exist, virtue would be a mere metaphysical concept.

11 *upon the major part* a preponderance.

15 *without a satire* without vituperation, calmly.

28–9 *trajection of a sensible species* perception transmitted to the mind by an emanation from a material object: simply, physical light.

31 *no man can judge...knows himself* a striking statement, but a *non sequitur*.

PAGE 79

4–5 *those which most do manifest...zeal* probably the Puritans.

9–10 *his own executioner* 'There are too many examples of men that have been their own executioners....But I do nothing upon myself, and yet am mine own executioner' (Donne, *Devotions*, XII). The hesitancy of Browne's 'as it were' contrasts sharply with the authority of Donne's assertion.

10 *Non occides* thou shalt not kill.

12 *Atropos* one of the three Fates, whose shears cut the thread of life spun by her sisters Lachesis and Clotho.

16–17 *saw that verified...of himself* saw the existence of death demonstrated, which previously he could not credit.

27 *those oblique expostulations* Job xi–xii.

PAGE 80

18–20 *I could never remember...and my friends* The greater cause— God, country and friends—always put self-interest out of mind.

COMMENTARY AND NOTES

23–5 *I do not find in myself...my blood* Browne lost his father at the age of eight. His mother may have alienated his affection by marrying again. He seems to have disliked his stepfather.

25 *fifth Commandment* 'Honour thy father and thy mother.'

28 *I never yet cast...woman* Browne married in 1641, after the composition of *Religio Medici* but before its publication.

31 *specifical unions* marriage, uniting those of one species.

33–4 *three most mystical unions* the Incarnation, the Holy Trinity, and friends who share a single identity.

PAGE 81

20–1 *in a competent degree affect all* will feel an appropriate degree of love for everyone.

32 *in my mirth and at a tavern* compare p. 91, ll. 17–18.

PAGE 82

9–10 *the story of the Italian* 'who, after he had inveigled his enemy to disclaim his faith for the redemption of his life, did presently poniard him to prevent repentance and assure his eternal death' (*Pseud. Epid.* VII, 19). A version of the same story figures in Nashe's *Unfortunate Traveller*, 1594; see McKerrow, II, 325–6.

29 *at sharps* with unbated swords; in deadly earnest.

31 *Battle of Lepanto* a naval battle between Christian and Turkish fleets fought on 7 October 1571, by which the Turkish threat to European civilisation was broken.

PAGE 83

17 *spintrian recreations* sexual perversions reported by Suetonius of the emperor Tiberius; *Tib.* 43.

18 *new and unheard-of stars* in consequence of Galileo's invention of the telescope, with which in 1610 he discovered the four satellites of Jupiter.

PAGE 84

1 *our great selves, the world* the macrocosm; the fully expanded entity of which man is the microcosm.

3 *particular discords* the mutual antipathy between fire and water,

and earth and air, by which the dangerous preponderance of any single element is prevented.

31 *the Pointers* the two stars of the Great Bear which indicate the position of the Pole Star.

PAGE 85

2 *simpled further than Cheapside* one of the main streets of London as Browne knew it; meaning before he knew anything of the world.

2–3 *heads of capacity* great minds.

7 *the riddle of the fishermen* 'He asking them what they had taken, they made him this enigmatical answer: That what they had taken, they had left behind them; and what they had not taken, they had with them: meaning that because they could take no fish they went to louse themselves; and all which they had taken, they had killed and left behind them, and that all which they had not taken they had with them in their clothes: and that Homer, being struck with a deep sadness because he could not interpret this, pined away and at last died' (Digby).

10–11 *flux and reflux of Euripus* 'That Aristotle drowned himself in Euripus, as despairing to resolve the cause of its reciprocation, or ebb and flow seven times a day...is generally believed amongst us' (*Pseud. Epid.* VII, 13).

16 *Peripatetics* philosophers of the school founded by Aristotle or based upon his work.

   Academics Platonist philosophers.

17 *Janus* a god with two faces looking in opposite directions.

33 *an accessory of our glorification* in heaven the blessed will possess knowledge as a condition of their state.

PAGE 86

7 *the rib and crooked piece* See Genesis ii. 22: 'crooked' in the double sense of the rib's bowed shape and of the moral perversity of Eve, which led to the Fall.

8 *procreate like trees* None the less, Browne's wife Dorothy Mileham bore him twelve children, four of whom survived their parents. Dorothy died three years after her husband.

23 *music of the spheres* harmonious sound supposedly produced by the motion of the celestial spheres in which each of the planets was fixed.

24-5 *though they give no sound...harmony* 'impossibile leggere questo passo senza ricordarne uno notissimo dello Shakespeare in *The Merchant of Venice*, v, i, 80-7' (Sanna, p. 248).

28 *against all church music* all Puritans of consequence were against organ and instrumental music in church, though they approved the singing of metrical psalms.

34 *a hieroglyphical...lesson* Browne's addiction to 'mystical types' again declares itself.

PAGE 87

3 *sensible fit* audible snatch.

7 *the soul is an harmony* compare *Urn Burial*, p. 37: 'the harmonical nature of the soul; which delivered from the body, went again to enjoy the primitive harmony of heaven'.

11 *first line of his story* 'Urbem Romam in principio reges habuere.'

13 *perfect hexameter* 'In qua me non infitior mediocriter esse.'

17 *fatal conjunctions* of stars, producing disasters on earth.

30 *three noble professions* medicine, the law, divinity.

PAGE 88

9 *stone* gall-stone.

12 *their precepts* spiritual advice of divines.

24 *Magnae virtutes nec minora vitia* great were his virtues, and no less his faults.

25 *may be inverted on the worst* the sense of the remark may be reversed when speaking of evil men: they have great virtues as well as vices.

31-2 *the greater balsams...powerful corrosives* a euphuistic assertion of doubtful truth.

PAGE 89

4-5 *contagion of commerce without me* infectious contact from outside.

6-7 *the man without a navel* Adam; since—as Browne argues in

Pseud. Epid. V, 5—never having been joined to a mother, Adam could not have possessed an umbilicus.

12 *Nunquam minus solus quam cum solus* never less alone than when solitary (Cicero, *De officiis*, III, i). Compare p. 13, ll. 1–3.

PAGE 90

4 *above Atlas' shoulders* In Greek mythology, Atlas bears the world on his back. 'Though he creepeth gently upon us at the first, yet he groweth a giant, an Atlas—to use his own expression—at the last' (Digby).

4–5 *though I seem...tiptoe in heaven* although I seem to stand on the earth, am reaching upwards into heaven: compare l. 31 below.

10–12 *though the number...my mind* my body can be measured, but my mind is measureless.

22 *Achilles* who, after his mother had dipped him in the river Styx, could be wounded only in the heel by which she had held him.
   fortune hath not one place to hit me an assertion condemned by Browne's remark on p. 22, l. 18.

22–3 *ruat coelum, fiat voluntas tua* thy will be done, though the heaven falls.

PAGE 91

15 *Scorpius* Scorpio, the eighth sign of the Zodiac. Browne's birthday was 19 October.

16 *planetary hour of Saturn* 'If Saturn be predominant in his nativity, and cause melancholy in his temperature, then he shall be very austere, sullen, churlish, black of colour, profound in his cogitations, full of cares, miseries, and discontents, sad and fearful, always silent, solitary' (Burton, *Anatomy of Melancholy*, I, iii, 3).

32 *Morpheus* god of sleep.

PAGE 92

8 *those spirits which are the house of life* 'Spirit is an airy substance subtle, stirring the powers of the body to perform their operations' (Elyot, *The Castle of Health*, 1541). The spirits were wasted by the labours of waking life.

14 *Lucan and Seneca* whom Nero allowed to choose what manner of death they should suffer.

COMMENTARY AND NOTES

14 *or to wake* whether to wake.

19 *to bedward* on going to bed.

24 *geometrical proportion* one involving an equality of geometrical ratio in its two parts; e.g. 1:3, 4:12. Problems involving multiplication were originally dealt with by geometry, and not by arithmetic.

PAGE 94

2-3 *some have held that snow is black* 'et Anaxagoras la disoit estre noire' (Montaigne, II, 222).

3 *that the earth moves* the Schoolmen held that the earth stood motionless at the centre of the planetary system, which revolved about it.

4 *all this is philosophy* fantastic notions; probably with an allusion to Cicero, *De divinatione*, II, 58: 'Nihil tam absurde dici potest, quod non dicatur ab aliquo philosophorum.'

6 *subterraneous idol* gold.

8-9 *its prepared substance* preparations of gold—*aurum potabile*—were used for medical purposes.

14 *will not allow us* will not admit that we can.

17 *example of the mite* Mark xii. 41-4.

28 *Peru* famous as a source of gold.

PAGE 95

14 *the prophecy of Christ* 'The poor ye shall have always with you' (marginal gloss). In fact neither John xii. 8 nor Matthew xvi. 11 uses the future tense.

29 *call to assize* examine.

31 *dumb-shows* mimed plays.

PAGE 96

6 *desire our ends* wish us dead.

11 *as Copernicus will have it* The Copernican theory of a heliocentric universe had been published in 1543, but had not yet won any general acceptance. Donne and Burton were among the many English writers who treated the theory with derision or amusement.

13   *All is vanity* Ecclesiastes ii. 11. Browne supposes that its author was Solomon.

15   *the Ideas of Plato* see note on p. 74, ll. 32–3.

16   *summum bonum* highest good.

16–17   *no such thing as his felicity* Aristotle concludes in the *Ethics*, x, 7, that the life of intellectual activity offers man the greatest happiness, but that such a life is too high for his attainment.

21   *a story out of Pliny* whose *Natural History* abounds in fables.

22   *Boccace* Giovanni Boccaccio, 1313–75, poet and novelist.
   Malizspini probably Celio Malespini, 1531–1609, author of *Duecento novelle* (Venice, 1609).

33   *Triuni Deo...aeternum* To God the Father, Son and Holy Ghost, be glory everlasting.

GLOSSARY

Asterisks denote words which the *Oxford English Dictionary* shows Browne to have introduced, or to have used in a new sense.

abrupts, interrupts, cuts short
abstracted, absent in mind, withdrawn
acception, act of accepting
accident, appearance
accidental, fortuitous, undesigned
accidentally, non-essentially
accrue, come by way of addition
actually, in reality
adjunct, addition
admiration, wonderment
adumbration, representation
adventurous, enterprising
adviso, counsel, suggestion
affect, love
affection, influence [17]; condition [42]; attribute [44]
alloy, quality, character
amazed, distracted, terrified
amphibious, having two lives
amphibium, creature with a double nature, able to live in water and on dry land
amphibology, amphiboly, ambiguous sentence, quibble
anatomy, anatomised body, corpse
angerly, angrily
anthropophagi, cannibals
antic, mountebank, grotesque clown
anticipatively, prejudicially
antimetathesis, inversion of the members of an antithesis
antinomy, conflict of authority

antipathy, natural contrariety or incompatibility
antiperistasis, force of contrast
apothegm, apophthegm, terse pointed saying
apparent, visible, palpable
appurtenance, accessory, necessary adjunct
argumentation, debate
artist, savant, man of science
ascendant, point of the ecliptic which at any moment is just rising above the eastern horizon
aspect, influence
asphaltic, like or containing asphalt, an inflammable bituminous substance
aspire, aspiration
asquint, obliquely
assassin, assassinate
assize, judgement
asunder, apart
auditory, lecture-room, philosophy school
authentic, authorised, accepted

basilisco, basilisk, large brass cannon
begat, called into being
beholding, indebted
belies, slanders
beneficence doing good
beneplacit, good pleasure
benevolous, auspicious

bespeak, engage beforehand, order

bethink, consider, suppose

bottom, vessel [4]; skein or ball of thread [52]

buckler, small round shield, means of defence

bungler, clumsy unskilful worker

burden, recurrent refrain

cadaverous, belonging to a corpse

caitif, mean

calcined, reduced to powder

canicular, pertaining to the dog-days during the hottest season, when creatures were likely to run mad

canker, disease

canon, decree of the Church

captious, carping, fault-finding

*carnalled, had intercourse with

*carnified, converted into flesh

catastrophe, conclusion, final event

catholicon, supposed drug to purge all humours, panacea

cento, patched garment

challenge, claim, assert title to

changeling, ugly or stupid child substituted for another by fairies

chiromancy, palmistry

chorography, topographical or geographical features of a region

circumstantial, adventitious

civility, organised purpose

climacter, climacteric year, critical period of a man's life

coition, sexual conjunction

colloquy, conversation

community, common character

commutative, pertaining to exchange or mutual dealings

compellation, title, style of address

competent, suitable, proper

complement, completeness

complemental, additional

*complexionally, temperamentally

complexioned, temperamentally disposed

compose, settle

compound, settle a liability

comprehend, include, take in

conceit, [sb.] imagining, fancy, notion; [vb.] suppose, imagine

conceited, imagined

concordance, agreement

concourse, concurrence in human action

*conformant, conforming, accordant

confound, confuse, alarm

conjoin, join together, come into conjunction

conjunction, intercourse, joining together

connive, wink at, disregard a fault

consist, stand firm, abide

consonant, agreeing or consistent with

consort, companion

constellated, predisposed by nature

construction, interpretation

contemn, despise

controvert, deny, contradict

conversation, way of life

conversion, turning

conveyance, means of transporting

convincible, capable of being proved false

corps, living body

corpulency, material substance, density

corrosive, destructive acid

cramb, distasteful repetition

cranie, cranium, brain-case

crasis, combination of elements or humours in an animal body

credit, reputation

cryptic, secret, occult

curious, intricate, delicate, minute

dastard, cow, terrify

decrepit, worn out, enfeebled

definitive, decisive, conclusive

deleterious, hurtful, injurious

delirium, mental disorder, madness

deliver, affirm, assert

demerit, quality deserving blame or punishment

demonstration, indubitable proof

deposition, deposing, putting down

derived, imparted, passed on

determinate, definitive, fixed in size and extent

dichotomy, division into two

difference, distinguish, differentiate

digestion, process of extracting soluble constituents from a substance by the action of heat

dilated, expanded, diffused

disavouched, disavowed, denied

discovered, disclosed

disparage, discredit, dishonour

dissembled, disguised, concealed

dissentaneous, at variance with, not in agreement

dissimilary, dissimilar, unlike

distinguish, separate, divide

distributive justice, distributing in shares proportionate to the deserts of each recipient

divinity, study of divine nature, theology; divine nature

donative, gift, bounty

**dorado, wealthy person

dormitive, narcotic drug

dormitory, sleeping-chamber, resting-place, cemetery

dotage, senility, second childhood

durst, dared

easy, not rigidly defined

economy, management, system of rules

ecstasy, rapture, frenzy of astonishment

ecstatic, affected by a trance or mystic absorption

edification, imparting of spiritual strength

edified, constructed, framed

efficacy, power of producing effects

**effront, free from bashfulness

elected, chosen by God for salvation

election, judgement

eleemosynary, one who lives on alms, beggar

elevation, devout exaltation of feeling

elixir, quintessential principle, soul

empirically, in the manner of a quack-doctor or empiric

empyreal, pertaining to the empyrean or highest heaven

engross, gain possession of, monopolise

enjoined, commanded

ephemerides, plural form of *ephemeris*, table of predictions, almanac

epicycle, small circle having its centre on the circumference of a larger one

epidemical, prevalent, universal

epitome, summary, condensation

equivocal, having the name but not the reality of a thing

essence, entity, being, substance

eternised, immortalised, perpetuated

ethnic, heathen, pagan

exaltation, process of refining or subliming a substance

excommunicate, excommunicated, cut off from religious rites

expansed, widely spread out

expatiate, allow to range, discuss freely

experimental, relating to experience

expostulation, protest, reproof

extemporary, without study or preparation

extirpate, root out

extravagant, irregular, fantastic, extreme

facetious, jocular, witty, gay

facetiously, waggishly, humorously

faculty, authority [19]; inherent power [35]

fallacy, unsoundness of argument, flaw

fallible, liable to be mistaken, unreliable

familist, member of a family or household

feasible, practical, probable, likely

feign, relate in fiction

feud, quarrel, contention

filament, minute fibre

filed, catalogued

fit, spell, short period

flux, change

forelaid, planned beforehand, prearranged

fraught, stored, supplied

fruitful, abundantly productive

fugitive, fleeting, inconstant

galliardise, gaiety, revelry

genealogy, descent

genius, natural ability, capacity

glome, ball or clue of yarn

gramercy, thank-you

graphical, pertaining to writing, consisting of letters

gratis, free

gravelled, nonplussed, puzzled

*grotesque, comically distorted or exaggerated figure

ground, reason

haggard, wild, unreclaimed

helix, spiral course

hellebore, purgative drug used in treatment of insanity

hermaphrodite, creature in whom both sexes are combined

hermetical, pertaining to occult philosophy

heterogeneous, diverse, completely unlike

hieroglyphic, emblematic figure

hieroglyphical, dealing with hieroglyphics or emblems

homicide, man-killing, murder

hoodwink, blindfold

horoscope, chart showing the disposition of the heavens which determines character and fortune

humour, [sb.] vein, inclination; [vb.] suit, fit

husband, cultivate, administer

husbandman, tiller or cultivator of soil, farmer

146

hypochondriac, deranged by melancholy

hypostasis, distinct substance, essence

idiosyncrasy, individual inclination, liking or aversion

immaterial, insubstantial

immured, enclosed, walled in

impassible, incapable of being injured

implicit, unquestioning, absolute

impregnate, permeated with an active principle

improperation, reproach, taunt

incommodity, disadvantage, inconvenience

inconsonant, not agreeing or harmonising with

incorrigible, depraved beyond correction or reform

incurvate, make crooked, bend

indifferency, ambiguity

indifferent, undetermined, neutral

indissoluble, insoluble, not to be ended

*inducible, capable of being inferred

inductive, reasoning from particular facts to general principles

infallible, certain, unfailing

inform, give material shape to

infusion, imparting of knowledge by divine force

ingenuous, liberally educated, cultured

inoculation, engrafting, budding

inorganity, being without organs

inquisition, enquiry, research

insensibly, imperceptibly, unconsciously

insolent, arrogant, disrespectful

interim, intervening time, meanwhile

inundation, flood

inveigle, draw away, beguile

inventory, catalogue, detailed account

invulnerable, safe from harm or attack

irradiation, shining ray of light, emanation

item, hint, intimation

jubilee, fiftieth year or anniversary

jurisdiction, power, authority

laudable, praiseworthy

laudanum, opiate, sleeping drug

lecture, reading, study

legacied, bequeathed

legerdemain, trickery, deception

lief, willingly, gladly

ligation, binding, suspension of faculties

limbo, abode of the damned who died before Christ's coming

limn, portray, depict

liquation, process of becoming liquid or molten

livery, allowance, stipend [57]; garments [75]

living, benefice, livelihood

luminary, light

magisterial, of a master-workman

magnify, praise, extol

maintain, uphold, support belief in

mannerliest, most decent, seemly

martyrology, register of martyrs

*materialled, brought into material form

meander, crooked path, winding

mechanic, manual labourer, common man

mendicant, beggar

mercenary, hireling

mere, absolute

meridian, greatest altitude of a star, point of highest development

metempsychosis, passing of the soul from one body to another at death

method, proper practice

mettle, quality, spirit

militant, one engaged in war

mind, take notice of, observe

miscall, revile, abuse

mischief, injury, hurt

miscreant, heretical, villainous, base

mislike, disapprove of

mistrust, suspicion

mock-show, ludicrous illusion

moderator, judge, arbiter

morality, moral lesson or significance

morosity, moroseness, gloomy unsociability

mortification, destruction of vital or active qualities

mortified, altered, broken up

motive, argument inducing belief

mutation, alteration, change

mutilate, imperfect

nativity, birth

naturality, naturalness

nauseous, offensive, highly unpleasant

negative, contradiction

nicety, subtlety, scrupulous particularity

noctambulo, sleep-walker

nonage, infancy, minority

nullity, nothingness

*numerical, particular, individual

object, present in discourse

oblique, morally perverse

omneity, allness, totality

opinioned, held a belief in

opprobrious, abusive, vituperative

optics, sight

orison, prayer

panoplia, panoply, full suit of armour

pantaloon, pantomime fool

paradoxical, given to paradox and fantastic ideas

parenthesis, explanation inserted into discourse

party, incidental part

pate, head, skull

patroned, acted as patron, patronised,

peccadillo, trifling offence, venial fault

pecuniary, monetary

peradventure, perchance, maybe

peregrination, travelling in foreign lands, journeying

peremptory, decisive, precluding discussion

periphrasis, amplification, larger expression

perspective, spy-glass, telescope

perusing, inspecting, scrutinising

petty, in miniature

phantasm, ghost, apparition

physic, medicine

physiognomy, art of judging character from the face

phytognomy, art of discovering properties of a plant from its appearance

piece, work of art, production, play

piece out, eke out, complete, enlarge

piece up, compose, make up

platonic, metaphysical

plebeian, pertaining to the common people

plunged, overwhelmed, perplexed

politic, diplomatic, temporising

politician, one skilled or experienced in state affairs

popularly, by the common people

portion, share, allowance

pose, assume, suppose

posy, motto, emblem

potion, draught of medicine, drink

prating, chattering, talking idly

predestinated, predetermined

preferred, advanced in status

pregnant, weighty, compelling, convincing

prejudicate, settled or decided beforehand, biased

premonition, forewarning

preordinate, fore-ordained, predestined

prerogative, special right or privilege

prescript, rule, law

pretend, profess, claim

privation, deprivation

privileged, invested with a privilege

probably, with probability

process, gist

prodigious, marvellous, enormous

profaned, vulgarised, made common

professed, professional

profound, immerse, plunge

progeny, descendants, offspring

prognostic, [sb.] prediction; [vb.] prognosticate, foretell

promiscuously, without distinction or discrimination

propagate, breed, spread

propense, having an inclination or propensity towards

proper, distinctive, characteristic

providence, divine direction or control

pucelage, virginity

punctual, definite, precise, exact

purgatory, place of spiritual purging and purification

purlieu, haunt, beat

quadrate, agree, correspond

quasi, kind of, resembling or simulating

queasy, sickly, upset, bilious

quotidian, daily, recurring every day

rabbin, rabbi, Jewish doctor of law

radical balsam, healthful preservative substance, believed to exist in all organic bodies

radical heat, warmth naturally inherent in living things, and a necessary condition of life

radical humour, natural moisture essential to life

railed, abused

really, in reality

*recompensive, that recompenses, repays

reflex, mental reflection, consideration

regiment, rule, order, mode of life

relish, partake of, carry tinge of

repine, complain, express discontent

reprehension, reprimand, censure

reprobate, [sb.] one lost in sin, depraved character; [vb.] condemn, reject

reserved, specially retained for some person

reserving, setting apart, holding back

resolution, resolute person

restraining, limiting, restricting

retaining to, kept in subjection to or dependence upon

retribute, give back, make return or requital

retrograde, going backward

reverberated, melted, subjected to the heat of a furnace

revolve, turn over (pages)

rhapsody, extravagant, disconnected or confused piece of writing

rhetoric, art of using language to persuade or influence others

rodomontado, vainglorious bragging or boasting

roundle, rung of a ladder

rub, obstacle, difficulty

rusticity, lack of breeding, clownish awkwardness

salve, make good, remedy, mend

scabbed, having skin-disease, ringworm

scandal, slander, imputation

scape, transgression, fault

scenical, fictitious, illusory

schism, division, breach of unity

scintillation, flash, spark

scruple, doubt, uncertainty

scrupulous, troubled with doubts, reluctant

secretary, one entrusted with secrets

secundine, placenta

semi-body, imperfect body

sensible, perceptible by the senses [23]; judicious, perceptive [58]; capable of being affected by objects of sensation [78]

sequestered, retired, withdrawn

shadowed, obscurely indicated, veiled

simpled, gathered simples, medicinal herbs

singularity, dissent from generally accepted opinion

sinister, unfortunate, disastrous

slough, outer skin, sheath

smattering, superficial knowledge

solary, sunlike

solecism, impropriety

solicitous, troubled, concerned about

sophism, faulty argument

sophistical, sophistic; in the manner of a sophist, employing specious arguments

sorites, syllogism, series of related propositions

sortilege, casting or drawing of lots

species, supposed emission or emanation from material objects by which they affect the senses *specifical*, specific; of characteristic kind, particular

spends, exhausts itself

spoiled, robbed, stripped

squares, harmonises, agrees

statist, politician, statesman

statute, statutory, recognised by statute

still, always, forever

stint, cessation, pause, stop

Stoic, one characterised by indifference to fortune

strain, passage of poetry

stupid, unfeeling, insensitive

styled, named, called

sublunary, earthly, temporal

subordinate, place in lower order, make secondary

subsist, exist as material substance or entity

subterraneous, subterannean, lying under the earth

superannuated, disqualified by age

supererogate, pay over and above

supposed, admitted, accepted

supputation, estimation, reckoning

swoon, fainting fit

synod, assembly, council

tables, backgammon

tacitly, by inference or implication

temper, temperament, disposition, nature

tenent, tenet, point of doctrine

tenor, drift, substance

tester, sixpence

textuary, regarded as authoritative

tincture, infusion, imparted quality

topography, outstanding features

traduction, transmission; something transmitted or derived

trajection, perception transmitted to the mind

translated, carried away, transported

*transpeciate, change into different form or species

trencher, wooden plate or serving-dish

trimmed, rigged

trope, figure of speech

tropical, metaphorical, figurative

tutelary, guardian, protecting

type, symbol, emblem; especially of Old Testament figures and events as prefiguring aspects of the new dispensation

typographer, printer

ubi, place, location

ubiquitary, which is everywhere at once

under-head, person of inferior intelligence

unspeakable, indescribable, inexpressible

urge, bring forward

*utinam, earnest wish, fervent desire

vacuity, vacuum

vegetation, property of growth

venery, pursuit of sexual pleasure

ventilation, breeze

venue, attack, assault

vespillo, one who buries victims of plague

viands, victuals, food

vicinity, nearness in space

GLOSSARY

vicissitude, mutation, uncertain changing

vitiosity, moral viciousness

vizard, mask

volatile, insubstantial, evanescent

vote, desire, ardent wish

warrantable, justifiable, authorised

without, outside

witty, intelligent, sharp-witted

wormed out, expelled, got rid of

worthy, famous man, national hero

wrench, violent turn or twist

zenith, highest point, culmination

INDEX OF PROPER NAMES

INDEX OF PROPER NAMES

Printed in the United States
By Bookmasters